Catastrophic Episodes in Earth History

Claude C. Albritton, Jr

CHAPMAN & HALL

London · Glasgow · Weinheim · New York · Tokyo · Melbourne · Madras

Published by Chapman & Hall, 2-6 Boundary Row, London SE1 8HN

Chapman & Hall, 2-6 Boundary Row, London SE1 8HN, UK

Blackie Academic & Professional, Wester Cleddens Road, Bishopbriggs, Glasgow G64 2NZ, UK

Chapman & Hall GmbH, Pappelallee 3, 69469 Weinheim, Germany

Chapman & Hall USA, One Penn Plaza, 41st Floor, New York, NY10119, USA

Chapman & Hall Japan, ITP - Japan, Kyowa Building, 3F, 2-2-1 Hirakawacho, Chiyoda-ku, Tokyo 102, Japan

Chapman & Hall Australia, Thomas Nelson Australia, 102 Dodds Street, South Melbourne, Victoria 3205, Australia

Chapman & Hall India, R. Seshadri, 32 Second Main Road, CIT East, Madras 600 035, India

First edition 1989
Reprinted 1995

Typeset in 12/13 Bembo by Thomson Press (India), New Delhi
Printed in Great Britain by Athenaeum Press Ltd, Gateshead
Tyne & Wear

ISBN 0 412 29200 9

A Catalogue record for this book is available from the British Library

Library of Congress Cataloging-in-Publication Data available

For my grandchildren, Lillian and Alexander Albritton, and Nicholas Scott

Contents

Acknowledgements

Documentation for the following text was provided by the Science Information Center at Southern Methodist University. There, with the aid of her computer, Reference Librarian Katheren A. Riddle was able to locate and obtain copies of books and articles found in only a few large collections, such as those of the US Geological Survey in Washington, D.C. Her colleague, geologist and bibliographer James Stephens, was especially helpful in tracking down references to the ongoing debate about the demise of the dinosaurs.

With regard to illustrations, I am indebted to Ms Carol A. Edwards of the Photographic Library of the US Geological Survey in Denver, Colorado, for searching her files for pertinent photographs, and for guiding me to published maps and geological structure – sections that are in the public domain. Dr Arthur Richards, Dr Ursula Marvin, and Dr Robert Dietz have been most generous in granting permission to reproduce photographs drawn from their private collections. Ms Jean Davis graciously provided a copy of the geological time scale copyrighted in 1983 by the Geological Society of America, and obtained permission for reproducing it here.

Acknowledgements

Series foreword

Year by year the Earth sciences grow more diverse, with an inevitable increase in the degree to which rampant specialization isolates the practitioners of an ever larger number of subfields. An increasing emphasis on sophisticated mathematics, physics and chemistry as well as the use of advanced technology have set up barriers often impenetrable to the uninitiated. Ironically, the potential value of many specialities for other, often non-contiguous ones has also increased. What is at the present time quiet, unseen work in a remote corner of our discipline, may tomorrow enhance, even revitalize some entirely different area.

The rising flood of research reports has drastically cut the time we have available for free reading. The enormous proliferation of journals expressly aimed at small, select audiences has raised the threshold of access to a large part of the literature so much that many of us are unable to cross it.

This, most would agree, is not only unfortunate but downright dangerous, limiting by sheer bulk of paper or difficulty of comprehension, the flow of information across the Earth sciences because, after all it is just one earth that we all study, and cross fertilization is the key to progress. If one knows where to obtain much needed data or inspiration, no effort is too great. It is when we remain unaware of its existence (perhaps even in the office next door) that stagnation soon sets in.

This series attempts to balance, at least to some degree, the growing deficit in the exchange of knowledge. The concise, modestly demanding books, thorough but easily read and referenced only to a level that permits more advanced pursuit will, we hope, introduce many of us to the varied interests and insights in the Earth of many others.

The series, of which this book forms a part, does not have a strict plan. The emergence and identification of timely subjects and the availability of thoughtful authors, guide more than design the list and order of topics. May they over the years break a path for us to new or little-known territories in the Earth sciences without doubting our intelligence, insulting our erudition or demanding excessive effort.

Tjeerd H. van Andel and Peter J. Smith
Series Editors

Preface

In the early 1930s when I took my first course in geology, I was taught that the continents and ocean basins had been in their present positions at least as far back in time as when marine life first became abundant. That view was embodied in something called the 'principle of the permanence of continents and ocean basins'. As a corollary, it then seemed reasonable to suppose that long cores into the ocean's bottom might disclose the history of marine life from the earliest times.

Although the continents were considered to be permanent fixtures, they were not supposed to have been static. Repeatedly they had moved up or down with respect to ocean basins. In the course of these bobbings the seas alternately flooded the continents and retreated from them. Sideways movements were disallowed, despite the wild speculations of a German meteorologist named Alfred Wegener.

The world view with which I was first indoctrinated was emphatically earth-bound. Granted that Meteor Crater and a few smaller holes in the ground had been formed by the impact and explosion of meteorites, extraterrestrial agencies were not considered to have played important roles in the ancient history of our planet.

So much has changed. Wegener's hypothesis of continental drift, after going through several mutations, emerged in the 1960s as the theory of plate tectonics. Evidence of seafloor spreading from the loci of oceanic ridges has proved that the ocean basins are not permanent global features. Cores of ocean-bottom sediments have failed to recover fossils older than around 150 million years, only about a fourth as old as strata on the continents containing the oldest fossils in abundance. Evidently the deep-ocean record of the earliest marine life has been destroyed by subduction along colliding plates.

Moreover, our older earth-bound view of earth history has been supplanted by something more cosmic. This change in perspective has precipitated an intellectual movement that has often been called neocatastrophism. Central to this new catastrophism is the idea that relatively brief episodes of rapid environmental changes over the globe have repeatedly interrupted the normal course of events. A related idea holds that these episodes were attended by crises in the history of life leading to mass extinctions of species.

The most recent and most thoroughly documented of the major biotic crises came near the close of the Cretaceous Period, around 65 million years ago. In 1980, Luis Alvarez and his associates boldly proposed that this episode of mass extinction was triggered by the impact and explosion of a large extraterrestrial body. The Alvarez hypothesis has been widely publicized in the mass media. No doubt this public interest has been aroused in part by the fact that the dinosaurs, most popular of fossils, were among the creatures that disappeared.

The impact of this hypothesis on scientific thought has been spectacular. Since 1979 several hundred papers have been published in respectable scientific journals, some supportive of the Alvarez speculation, others attacking it, and still others offering alternative explanations for the Cretaceous and earlier major episodes of mass extinctions. One observer, noting the claims and counter-claims in what has turned out to be a spirited controversy among the theorists, declared that we are witnessing 'science gone mad'. Most others, however, appear to thrive on the excitement that has been provoked; and they attribute any seeming madness to normal derangements attending an intellectual revolution getting underway.

A remarkable feature of the present controversy is that representatives of nearly all branches of natural science have been drawn into it. In addition to geologists and paleontologists, who heretofore have claimed the patent on the mass extinction puzzle, astronomers, physicists and geophysicists, chemists, botanists and zoologists are now having their say. Approaches to a common set of problems from widely different specialist perspectives admittedly have sometimes led to a confusion of tongues, but not yet to madness.

Regardless whether the impact and explosion of an extraterrestrial body was the cause of the terminal Cretaceous extinctions, it now seems clear that the Earth has been bombarded by such objects at intervals throughout its long history. Meteorites, asteroids and comets course about in the solar system and have left their marks on the moon and the other terrestrial planets; and there is no reason to suppose that our planet has escaped being a target.

Craters formed by meteoritic impact and explosion are the rarest of ephemeral land-forms on Earth. But the effects of explosion there manifest in the shattering, deformation, and metamorphism of the target rocks are also found in many of the hundred or so anomalous structures scattered over the continents and ranging in age from Precambrian into the Tertiary. These structures have variously been called 'meteorite scars', 'astroblemes', 'cryptovolcanic structures' and 'cryptoexplosion structures'. None shows evidence of volcanic activity, and the conclusion that at least some were formed by impact and explosion seems inescapable.

Approximately the latter two-thirds of the text treats the flux in ideas leading to the emergence of the new catastrophism and to the criticisms leveled against it. Introductory chapters cover earlier modes of catastrophism going back to ancient traditions of universal floods.

In preparing this review, I have been asked simply to report and not to 'take sides' in arguments pro and con. So at the outset I should confess to two biases. I was one of those early advocates of a meteoritic origin for certain cryptoexplosion structures, as opposed to the then current view that they were formed by explosive volcanic activity. I have found no reason to change that opinion. Secondly, I have a strong dislike for 'isms', for reasons expressed in the concluding paragraphs of the text.

Claude Albritton
Dallas, Texas
Spring, 1988

1 Historical and legendary disasters

1.1 NATURAL DISASTERS OF HISTORICAL RECORD

During the period of 34 years ending in 1980, natural agencies of destruction accounted for the deaths of an estimated 1.2 million persons. Of these mortalities, 37% have been attributed to earthquakes, as compared with 16% to floods (Shah, 1983; Murty, 1986).

Over the long term, however, floods would appear to have been operationally the most lethal and geographically the most widely distributed. A listing of historic disasters over the world from 1642 to 1973 indicates that upwards of 4.9 million persons perished in floodwaters. The most catastrophic floods of record occurred along the Yellow River (Huang Ho) of China. During the flood of August 1931, upwards of 3.5 million lives were lost, as compared with an estimated 900 000 in the flood of 1887 (Whittow, 1979).

1.2 LEGENDARY ACCOUNTS OF FLOODS

Stories of floods reputed to have been even more destructive than those of the Yellow River are incorporated in various religious traditions. The biblical account of Noah's flood is the catastrophe of particular interest in the present context, because it left its mark on scientific thought during the time when geology was emerging as a science.

The account of Noah and the universal flood appears to have been an adaptation of an earlier version contained in the Babylonian Gilgamesh Epic. Clay tablets recovered from the library of Ashurbanipal (668–633 BC) at Nineveh tell the adventures of Gilgamesh (Sanders, 1976).

Gilgamesh was a demigod – son of a goddess and a human father. From his mother he inherited his beauty, strength and restless nature; from his father, his mortality. The tragedy of the epic revolves around the conflict between godly desires and the mortality of humankind. At one point Gilgamesh relates the following story about a universal flood:

In the days before the flood, the world teemed with people. With multiplication of the population, the noise level increased to the point that Enlil (God of the Earth, wind and universal air) couldn't sleep. So Enlil persuaded the other gods that mankind should be exterminated by drowning. But Ea, God of wisdom and one of the creators of humans, warned a man named Utnapishtim of this plot by means of a dream. Utnapishtim dreamt that he had received a divine command to tear down his house and build a boat.

With the aid of family members and craftsmen, the boat was built in six days. Square in plan, the craft had six decks. Onto it Utnapishtim loaded his family and kinfolk, craftsmen, and wild and tame beasts, together with provisions to sustain them all.

Soon afterward, rain began to fall in torrents. Water also gushed upward from the Abyss. Even the gods were terrified at the sight of the mounting flood. For six days and six nights the tempest raged. On the seventh day the storm subsided and the waters grew calm. In the distance Utnapishtim could see Mount Nisir. The boat sailed there and was grounded on that mountain for six days. On the seventh day the captain loosed a dove. She flew away, but could find no resting place and so returned. Then Utnapishtim loosed a swallow, with the same result. Then he loosed a raven. The raven, seeing that the waters had retreated, cawed, flew about and did not return to the boat.

Utnapishtim then went to the mountain top, prepared a sacrifice to the gods and poured out a libation. Smelling the sweet savor, the gods flocked like flies over the sacrifice. Enlil was the last to arrive, and when he saw that humans had survived the flood he was wroth. But Ea reproached Enlil and explained that he had not revealed the secret of the gods: Utnapishtim had divined it in a dream. Impressed and appeased, Enlil blessed Utnapishtim and his wife, and sent them to live as immortals far away at the mouth of the rivers (presumably the Tigris and Euphrates).

Older versions of the legend show that Ashurbanipal's version was based on a Sumerian original that goes back to around 3400 BC

(Vitaliano, 1973). Geologists studying alluvial deposits in the Tigris–Euphrates drainage have found evidence of a major flood dating from the fourth millenium BC. Thus the several mythic versions of the deluge that originated in Mesopotamia may trace back to a natural catastrophe, dimly remembered and encrusted with folklore.

1.3 A NATURALISTIC ACCOUNT OF THE DELUGE FROM THE 17TH CENTURY

That the catastrophic Noachian deluge resulted from natural as opposed to supernatural causes was a thesis of *The Sacred Theory of the Earth*, a popular work by the Reverend Thomas Burnet, first issued as a Latin edition in 1681. At the urging of Charles II, Burnet prepared an edition in English which was published three years later. *Sacred Theory* has been called the most popular geological treatise to come out of the 17th century. A sixth edition appeared in 1726, and the book was reprinted in 1965. In brief, Burnet's theory runs as follows (Figure 1.1):

The Earth originated from a chaos – 'a fluid Mass of all sorts of little parts and particles of matter, mixt together, and floating with confusion, one with another'. The particles drew together to form a perfect sphere with a smooth surface. In the process the heavier particles sank to the center, and the lighter ones rose to the top so that there were concentric shells of liquid and air around a solid core. In addition to ordinary water, the liquid shell contained an outer layer of oily fluid. Dust settled into this outer shell, soaking up the oil until it became a firm and fertile envelope above the layer of ordinary water.

The first generations of mankind inhabited this smooth paradisical Earth. Because the Earth's axis was then upright and not tilted, there were no seasons – 'every Season was a Seed-time to Nature, and every Season a Harvest'. Moisture drawn up from the waters below by the sun's heat settled as dew or fell in light life-sustaining showers.

Eventually the heat of the sun caused the fertile shell to dry and crack. The waters below boiled, vaporized, and finally exploded. Thus were 'the fountains of the Great Abysse' loosed, causing all parts of the Earth to be flooded. As the agitation of the waters subsided, they drained back into the lower places. Upward projecting fragments of the paradisical world were left standing as

Ἐγώ εἰμι τὸ Α καὶ τὸ Ω.
Τετέλεσται.
Ἀπὸ καταβολῆς κόσμου.
The
Sacred Theory
of the
EARTH.

continents and islands. Thus the world we live in is one great ruin, 'a broken and confus'd heap of bodies, plac'd in no order to one another'.

According to Burnet's philosophy, God created Heaven and Earth and then let Nature take her course. He was contemptuous of the proposition that torrential rains falling forty consecutive days and nights could flood the entire Earth. According to his calculations, a body of water about eight times that of the present oceans would be required to produce a flood that would top the highest mountains. Those who suppose that God was a magician who created this superabundance of waters and then caused them to disappear after they had done their work 'make very bold with the Deity', Burnet declared. No, Divine Providence decreed that the course of Nature should be exact and regular. As maker of the celestial clock, God could not only design a machine that would strike the hours regularly, but could also fashion one that would fall apart catastrophically at some appointed time, e.g. at the time of the deluge.

Burnet's heterodox views have been praised by Steele, Addison and Wordsworth; and ridiculed by Swift, Pope and Gay. In William King's ballad, 'The Battle Royal', Burnet is cast as an atheist, who proclaims 'that all the books of Moses are nothing but supposes'. In the end, Burnet lost his battle with the fundamentalist Christians, and with that his chance to be named Archbishop of Canterbury.

The Burnet controversy did, however, raise popular interest in the history of the Earth. Also, it challenged others to come up with better theories. A major problem of the times was to account for the opening of floodwaters from the Abyss. To solve that puzzle, William Whiston, who succeeded Isaac Newton at Cambridge, proposed in 1696 that the close approach to the Earth by a comet had

---

Figure 1.1 Frontispiece of Burnet's *Sacred Theory*. The seven spheres illustrate Burnet's ideas on how the Earth arrived as its present condition and on what will happen to it hereafter. Beginning at the upper right and proceeding clockwise, we see in succession the primordial Earth, the smooth paradisical Earth, the Earth enveloped in Noah's flood (with the ark shown riding the waves), the Earth in its present ruinous state, the Earth in flames, the Earth restored to a paradisical condition, and the Earth finally transformed to a star.

caused great rains to fall and the Abyss to loose its subterranean waters.

1.4 THE USSHER CHRONOLOGY

In 1654 James Ussher, Irish scholar and Archbishop of Armagh, had devised a system of Scriptural chronology which placed the date of Creation at 4004 BC. That became the date imprinted in many editions of the Bible, and it was probably the basis for Burnet's assumption that the Earth was created from the chaos around 4000 BC. In any case the presupposition that the Earth was only a few thousand years old led to a sort of obligatory catastrophism, manifest in several seminal works of the latter 17th century.

2

Obligatory catastrophism of the latter 17th century

2.1 CONSTRAINTS ON THEORIZING BASED ON BIBLICAL CHRONOLOGY

Burnet's naturalistic predecessors

Burnet may have been the premier popularizer of geological ideas during the second half of the 17th century. He was not however the first nor indeed the most persuasive advocate of the proposition that major changes in the Earth's configuration since its creation have been due to natural causes. Burnet gleaned his information mainly from documents, ancient as well as contemporary.

Four years before the Latin edition of *Sacred Theory* was issued, Robert Hooke (1635–1703) and Nicolaus Steno (1638–1686) had independently enunciated principles that today are regarded as essential to historical geology. Both based their ideas mainly on observations of rocks and fossils at the outcrop. Both were constrained in their thinking by the prevailing idea that the Earth was only a few thousand years old. Hence both conceived that some of the major changes in the planet's configuration during that brief past must have proceeded in jigtime.

2.2 STENO'S PRODROMUS

Steno and the shark

Steno's views on historical geology are set forth in a classical work published in 1669 under a Latin title which in English would read *Prodromus of Nicolaus Steno's dissertation concerning a solid body enclosed*

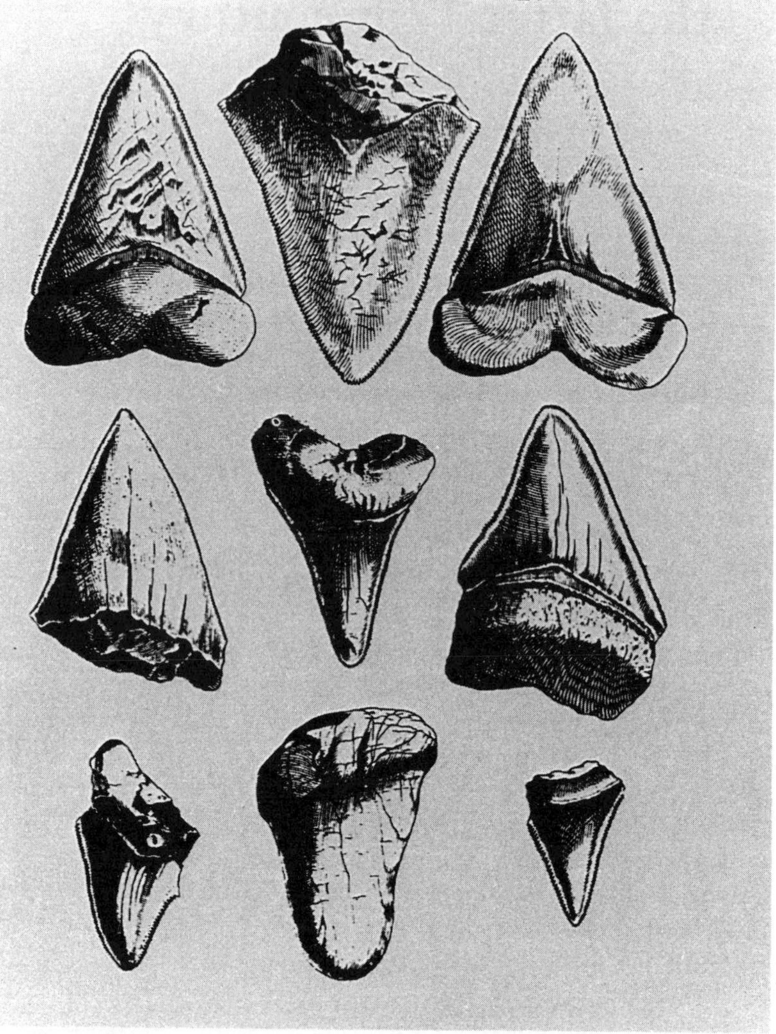

Figure 2.1 Glossopetrae. Plate prepared by the papal physician and naturalist Michele Mercati (1514–1593) and reproduced in Steno's shark's-head treatise.

by process of nature within a solid (English translation of text, 1968). A 'prodromus' is a preliminary discourse, but Steno never got around to writing the larger text of which this was supposed to be only an abstract. Here he was attempting to explain how one solid (a fossil) came to be enclosed by another solid (a rock). In the course of his arguments for the organic origin of fossils, he went on to formulate three principles important to historical geology.

Steno's interests in fossils and rocks came as a result of a curious coincidence. After completing work for the MD degree at the University of Leiden, he went to Paris where he conducted research on muscles and embryos. Then in 1665 he set out for Italy, where he was fortunate enough to gain the favor of Grand Duke Ferdinand II of Tuscany. In the fall of the following year, fishers in the Ligurian Sea caught a great white shark and dragged it ashore near Leghorn. News of that spectacular catch spread to the Medici Court in Florence, whereupon Ferdinand ordered the shark's head be cut off and fetched to Steno for dissection. When Steno first viewed the battery of sharp triangular teeth lining the jaws, he was immediately reminded of objects known from antiquity as tonguestones (*glosso-petrae*) (Figure 2.1).

Maltese tonguestones

Though found in many parts of the world, these objects are especially abundant in rocks of the Maltese Islands. They are triangular in plan, their thick bases tapering to narrowly rounded or pointed ends along sides with serrate edges. The faces are commonly enameled and shining. Lengths range to upward of 4 in (10 cm). Overall, these objects bear a fanciful resemblance to tongues that have turned to stone.

In his influential *Natural History*, composed during the 1st century AD, Pliny the Elder proposed that the tonguestones had fallen from the sky during eclipses of the moon (Plinius, 1601). According to a Maltese tradition, these objects were miraculously created following the landing of Paul the Apostle after his ship was wrecked near the bay that today bears his name. Bitten by a viper, Paul was angered but not poisoned. He placed a curse on all snakes thereabouts, whereupon their teeth turned to stone. A third tradition, widely held when Steno wrote his *Prodromus*, considered tonguestones simply as 'sports of nature', as minerals grown in the

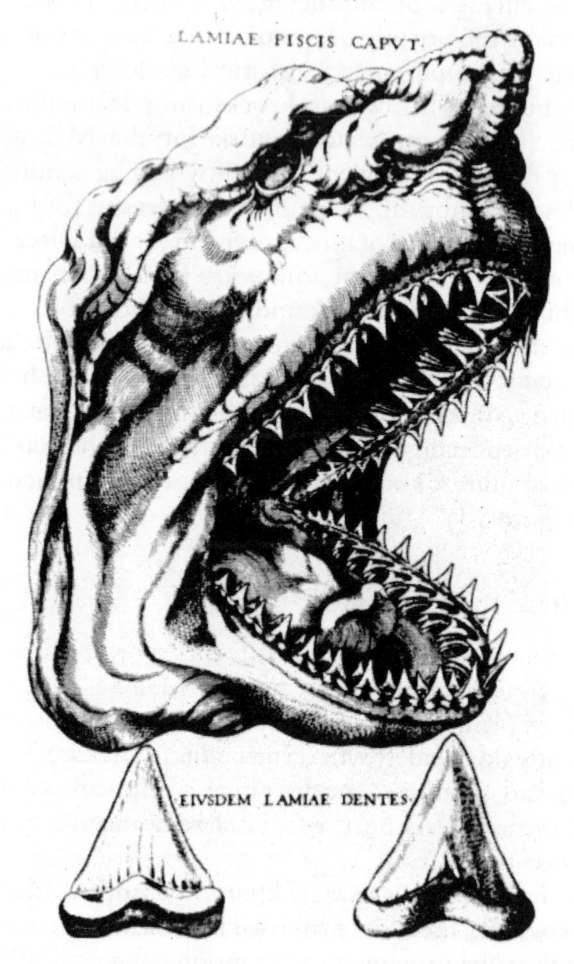

Figure 2.2 Head and teeth of a modern shark. From an imprint of a plate by Mercati, reproduced in Steno's shark's-head treatise.

ground, bearing only a coincidental resemblance to tongues or teeth (Figure 2.2).

Convinced from the outset that these objects must be teeth and not the products of a frolicsome Nature, Steno cautiously defended his hypothesis with a chain of conjectures. To begin, objects resembling the parts of aquatic organisms, seashells as well as tonguestones, do not seem to be growing in the ground today. On the contrary, they appear to be in the process of destruction when embedded in soil. Wherever these objects are concentrated on the surface, that must have been due to removal of the matrix by rainwash before they could be destroyed in the soil. And, if grown in the ground, they must have accommodated themselves to whatever cracks and openings were available. But whether in hard or soft ground, objects of the same kind maintain their characteristic shapes.

Steno could find no objection to the proposition that the matrix enclosing the objects in question was originally a sediment accumulated during the time when the present land was flooded. Sudden changes in sea-level could have been caused by violent earthquakes or explosive submarine emanations. Besides, Scripture records two instances of universal flooding. With respect to Malta, he suggested that 'formerly when this place was submerged in the sea it was the haunt of sharks, whose teeth in times past were buried in the muddy sea-bed'.

Stratigraphic principles

Steno observed that wherever the objects resembling parts of organisms are abundant, they occur in sequences of layered rocks. Having established to his satisfaction that the matrix containing these objects, no matter how hard at present, originated as layers of sediment deposited in water, he went on to argue that sequences of strata must have accumulated layer by layer and not all at once. Therefore, 'at the time when any given stratum was being formed, all the matter resting upon it was fluid; and therefore at the time when the lowest stratum was being formed, none of the upper strata existed'. Or, in modern terms, this principle of superposition of strata states that in a sequence of sedimentary strata as originally deposited, any stratum is younger than the one upon which it rests and older than the one that rests upon it.

Furthermore, Steno reasoned that during the accumulation of a stack of strata, while the older ones would reflect the irregularities of the bottom, as sedimentation progressed these irregularities would be ironed out so that the younger beds would be flat-lying. He framed this principle of the initial horizontality of strata as follows: 'Strata either perpendicular to the horizon or inclined toward it were at one time parallel to the horizon'.

Finally, he noted that when strata were first deposited they must have been entire. Thus when we see the bared sides of strata exposed in croppings, we may infer that something has been removed since the strata formed – either some original barrier that prevented the spread of sediment, or some parts of the strata themselves. This is the principle of stratal continuity.

Geological history of Tuscany

Having enunciated these three principles, Steno applied them to a reconstruction of geological history. In the area around Florence which Steno knew so well, there are two principal sets of strata. A sequence of relatively soft flat-lying beds rests upon and abuts against a sequence of harder tilted sedimentary rocks that form the mountains thereabouts. By the principle of superposition, therefore, the softer set must be the younger of the two. Applying the principle of initial horizontality, one can infer that at the time of origin the tilted strata also lay flat.

Thus Steno was led to conclude that Tuscany had twice been flooded by the sea. During the first episode the older strata were deposited. Then they were tilted. Steno speculated that this deformation was due to collapse of the roof above a gigantic cavity that had developed in the older strata after their formation. The younger and softer strata were deposited in the course of a second invasion by the sea (Figure 2.3).

Rationalizing this account with Scripture, Steno suggested that the second invasion by the sea bears witness to Noah's flood. The older set of strata he attributed to deposition in the universal ocean after the second day of Creation. Allowing for a date of Creation around 4000 BC, the interval between the two floodings would amount only to about 1650 years.

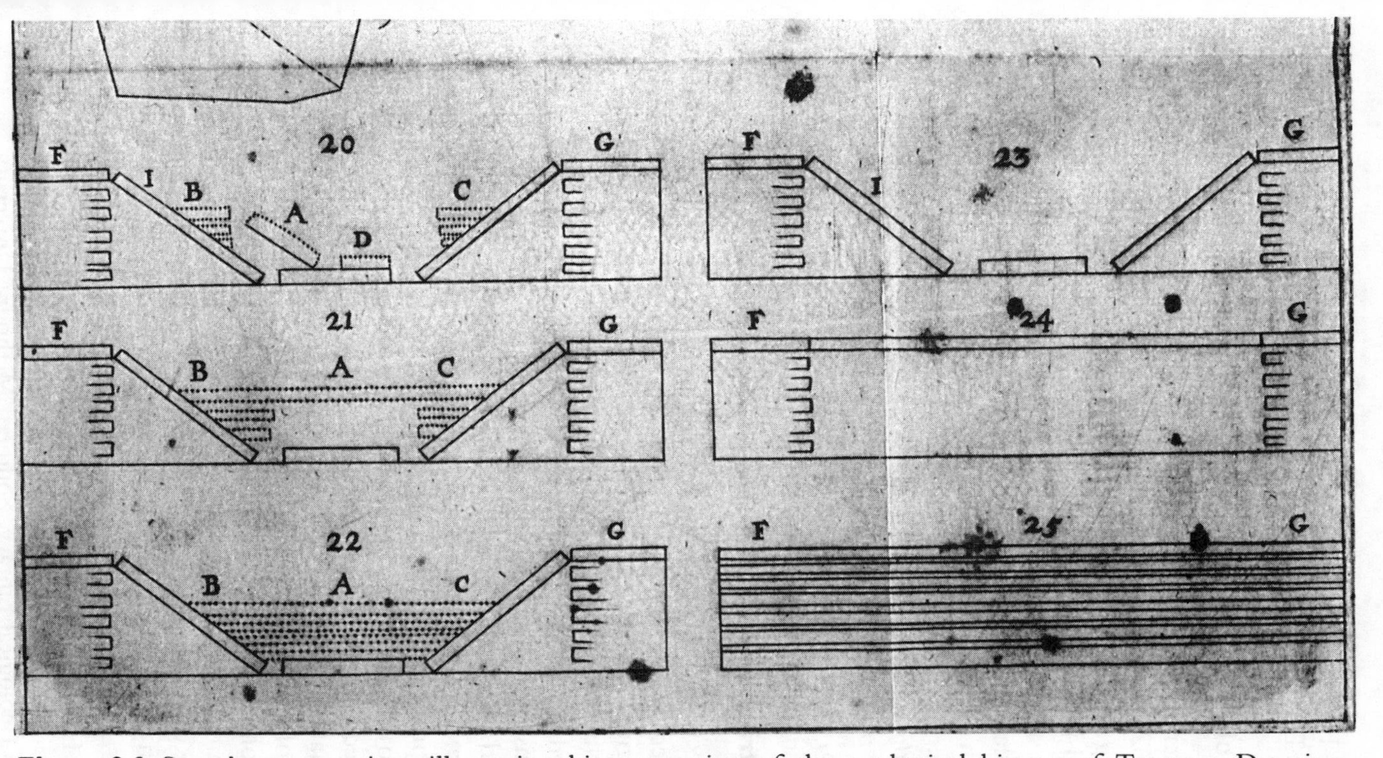

Figure 2.3 Steno's cross-sections illustrating his conception of the geological history of Tuscany. Drawings numbered 25 and 22 respectively show the deposition of the older and younger sets of sedimentary rocks. Drawings 24 and 21 show gigantic subterranean cavities opened in the two sets. Drawings 23 and 20 show the deformation of strata produced by the collapse of roofs over the cavities. From an original copy of the *Prodromus* in the DeGolyer Western Collection, Southern Methodist University.

2.3 HOOKE'S VIEWS ON FOSSILS, FLOODS AND EARTHQUAKES

Hooke's connections with the Royal Society of London

Soon after the Royal Society was established by charter of Charles II in 1662, Hooke was appointed Curator of Experiments. The following year he was elected a Fellow of the Society. Thereafter he delivered before the members many lectures on a wide variety of scientific subjects. Most of his discourses that related to geology and paleontology were made available to the general public through a collection of his works issued in 1705 under the lengthy title *Lectures and discourses of earthquakes and subterraneous eruptions, explicating the causes of the rugged and uneven face of the Earth; and what reasons may be given for the frequent finding of shells and other sea and land petrified substances, scattered over the whole terrestrial superficies.*

Petrifactions

The objects known today as fossils Hooke called petrifactions. He recognized two kinds. One group includes objects that preserve their original substances, such as bone, teeth, shells, and wood. The other includes objects made of mineral and earthy substances which have filled up and taken the shapes of the original shells, bones, fruit, etc. In this second group, animal or vegetable substances have been converted into stone by one of several processes. Their pores may have been filled up with some petrifying substance; or they may leave impressions on the surrounding matrix, 'such as heated wax affords to the seal'; or they may be casts made of materials later introduced into these natural molds.

By whatever process they were formed, petrifactions resembling the parts of animals and plants are widely distributed over the world, Hooke pointed out. They have been collected from hard rocks in the highest mountains as well as in the deepest mines.

Hooke was contemptuous of the idea that the petrifactions are no more than artifacts of a sportive Nature idly mocking herself. If these

---

Figure 2.4 Hooke's illustrations of ammonites. Reproduction of Plate 1 in *Lectures and discourses of earthquakes*, from an original copy in the DeGolyer Geological Library, Southern Methodist University.

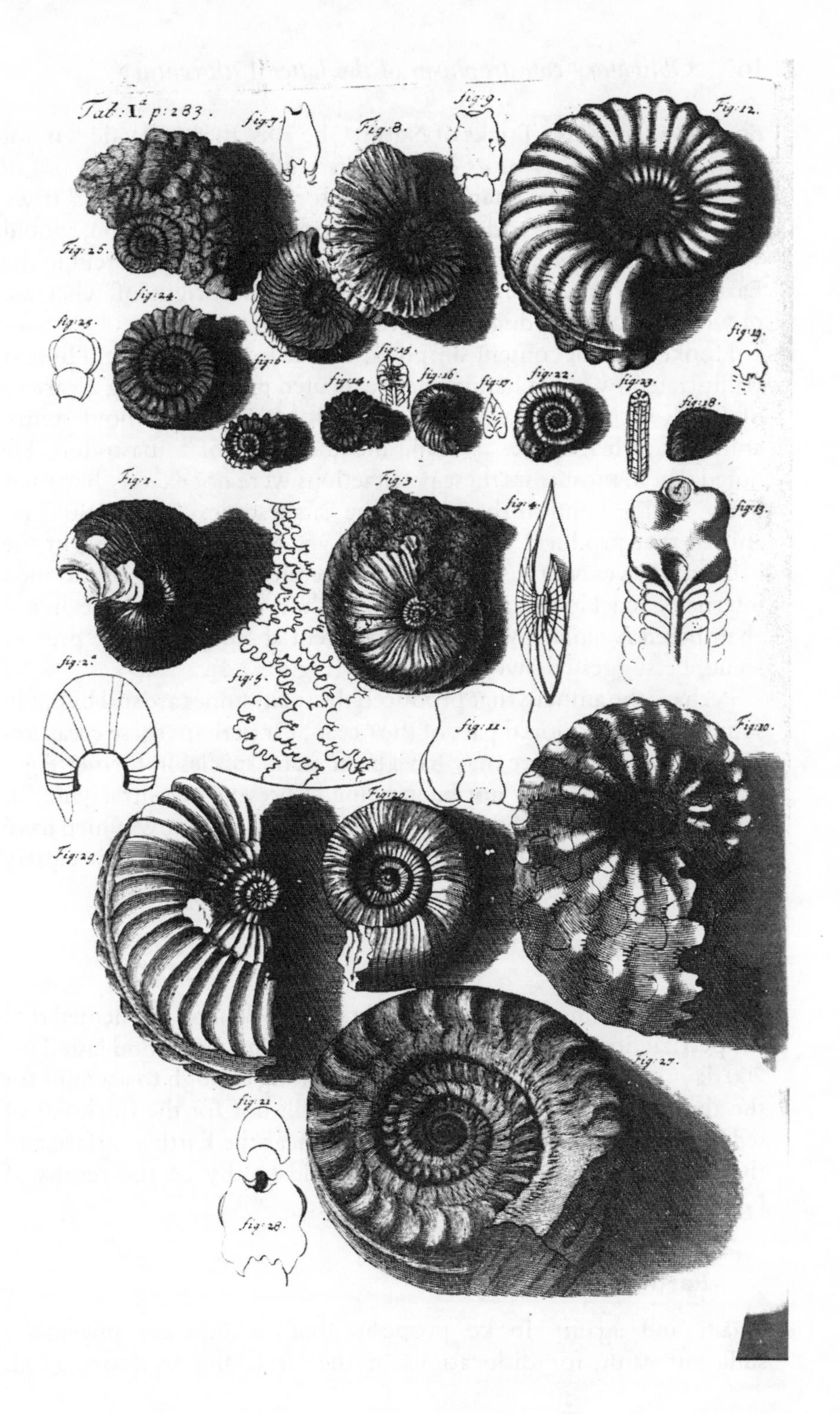
Tab: I. p: 283.
fig: 7
fig: 9
Fig: 8
Fig: 12.
Fig: 25.
fig: 24.
fig: 25
fig: 6
fig: 19
fig: 14
fig: 15
fig: 16
fig: 17
fig: 22
fig: 23
fig: 18
Fig: 1
Fig: 3
fig: 4
fig: 13
fig: 2
fig: 5
fig: 11
Fig: 10
Fig: 20
Fig: 29
Fig: 27
fig: 21
fig: 28

objects 'be the apish Tricks of Nature,' he asks, then 'why does it not imitate several other of its own works?' Why do we not dig out of mines 'everlasting vegetables, as Grass for instance, or Roses?' If we should turn up coins and urns in the course of an excavation, would we say that these were formed by some plastic faculty inherent in the Earth? If not, then why question the organic origin of what are obviously sea shells dug out of the ground?

Hooke was not content simply to verbalize his argument. His text is illustrated by seven handsomely executed plates showing a variety of fossils including clams, snails, corals, echinoids, crinoid stems, ammonites, belemnites, a crab, and the tooth of a mastodon. He noted that in most cases these petrifactions were not exactly like their living analogs; and in the case of the 'snakestones' (ammonites) no animal that produces these shells has yet been found living in the oceans. However, the fact that these coiled shells are partitioned internally led Hooke to suggest that they were kin to the living chambered *Nautilus*. As for their apparent absence from the present scene, he suggested two possibilities (Figure 2.4).

Perhaps the animals that produced the snakestones are still living in some as yet unexplored part of the ocean. Or perhaps these creatures are extinct, since 'there may have been many species in former Ages of the world that may not be in being at present'. Contrariwise 'tis not unlikely also that there may be divers new kinds now which have not been from the beginning'. Here Hooke was pretty clearly thinking along evolutionary lines (Rossiter, 1935).

Noah's flood

Hooke was willing to admit the reality of the deluge, but denied that the petrifactions provided evidence for it. Even if the flood lasted for 200 days, that time would not have been long enough to account for the abundance of fully developed sea shells nor for the thickness of sediments covering them. Fossils buried near the Earth's surface and those found at great depths below could hardly be the results of Noah's flood or any other single deluge.

Earthquakes

Again and again Hooke proposes that earthquakes provide a sufficient cause for dislocations of the lands and seafloors. High mountains, such as the Alps, have evidently been thrown up in the course of some very great quake. Probably the greatest part of the irregularity of the Earth's surface, he speculated, has proceeded from the 'subversion and tumbling thereof by some preceding Earthquakes'. In the course of these sudden dislocations, 'many Parts which have been Sea are now Land', which accounts for the fact that petrifactions of marine life are widely dispersed over the continents. These remains clearly indicate that most if not all parts of the British Isles have had fish swimming over them in former times, he concluded.

Age of the Earth

Hooke frequently refers to 'past ages' or 'many ages' without attaching numbers in years. Internal evidence indicates, however, that he accepted the biblical chronology. At one point he analyses the Atlantis story as told in Plato's *Timaeus* to determine whether it be factual or not. Referring to the time when the mythical continent was supposed to have disappeared, he concludes 'I confess that the account of the nine thousand years is argument enough to make the whole history to be suspected as a Fiction'.

Obligatory catastrophism appraised

Both Steno and Hooke appealed mainly to natural causes to account for past changes in the Earth's configuration. Constrained by contemporary calculations for the age of the Earth based on biblical accounts, these changes had to proceed rapidly, as in the course of a great earthquake or the sudden collapse of the roof over a huge cavern. None the less, they offered convincing proof that fossils are remains of organisms and not sports of Nature. Steno's stratigraphic principles, moreover, provided the basis for inferring sequences of past events in layered rocks. Application of these principles to the thick accumulations of sedimentary rocks in Europe and elsewhere led to far more generous estimates of the Earth's age framed in the 18th century. Catastrophism would survive in different guises, but would no longer be foreordained.

3 The antiquity of the Earth as perceived in Neptunist and Plutonist theories of the 18th century

3.1 NEPTUNIAN THEORIES

Basic assumption

The theories of the Earth most widely held during the 18th century have been called Neptunian, because they held in common the idea that in the early stages of its development the planet was covered by a universal ocean. With the secular lowering of sea-level, the present continents have emerged.

Otherwise these theories varied in detail. Some theorists held that existing land-forms were molded by oceanic currents prior to their emergence. Others allowed for modification of emerged land-forms by stream erosion during late stages of the Earth's history. The retreat of the sea was variously conceived as gradual and uninterrupted, or as episodic.

The concepts developed by Benoît de Maillet (1656–1738), Georges-Louis Leclerc, Comte de Buffon (1707–1788), and Abraham Gottlob Werner (1750–1817) illustrate well the similarities and differences among Neptunian theories.

De Maillet's theory

Born of a noble family of Lorraine, de Maillet was appointed consul general to Egypt aged 35. Thereafter he served successively as consul at Livorno, and inspector of French establishments in the Levant and along the Barbary Coast. On the basis of his geological observations around the Mediterranean shores, he began formulating his theory in 1692. Expanded versions circulated in manuscript around French intellectual circles after 1720. The first published version, however, did not appear until ten years after de Maillet's death. It was entitled *Telliamed: or Conversations between an Indian philosopher and a French missionary on the diminution of the sea.* (Telliamed is de Maillet spelled backwards.) Despite the fact that the fearful editor had omitted or modified parts of the original he thought might offend the pious, de Maillet's materialistic message was clear enough to be denounced by Voltaire and others less famous.

As controversy concerning *Telliamed* increased, so did sales. Before the end of the century two additional French editions were issued, and translations had been published in England and America. In 1968, Albert Carozzi's definitive English translation and critical commentary established that this was not a work of the unbridled imagination, but rather was a serious effort to reconstruct the ancient history of the Earth based mainly on observations made in the field.

At an early stage in its development, de Maillet theorized that the solid Earth was covered by a universal ocean. Evaporation of water into outer space has caused sea-level to fall. Meanwhile, powerful submarine currents were heaping up sediments along the ocean's floor to form seamountains and seavalleys. With continued diminution of the ocean, these emerged to form the primitive mountains. Along those ancient coast lines, waves and currents eroded the borders of these mountains and deposited the waste offshore. In turn, those sedimentary deposits emerged to form a series of secondary mountains, the most recent of which are being subjected to the marine erosion we witness today.

According to de Maillet, the seeds of organisms began to germinate in warm shallow waters at the stage when the primitive mountains were about to emerge. Seaweed, shellfish, and fish increased in number and kind, and their remains became entombed in sediments swept offshore to lay the foundations of the secondary mountains.

Thus all life on Earth originated in the sea. As the continents increased in size with the diminution of the ocean, marine life invaded the lands. Seaweed gave rise to shrubs and trees. Animals that had crept about on the seafloor transformed into animals that walked on the land. Elephant seals, de Maillet speculated, may have turned into elephants, flying fish into birds. Tritons and mermaids may have been the ancestors of mankind.

In the course of his travels, de Maillet had observed that at Carthage a seaside fortress has basement openings apparently designed to admit sea water. The base of these openings stands 5–6 ft (1.5–1.8 m) above present sea-level; thus in the course of about 2000 years sea-level must have dropped by an average of 3–3.6 in (7.6–9 cm) per century. Similar situations at Acre and Alexandria supported that conclusion. On the assumption that the diminution of the ocean had progressed at a steady pace through time, de Maillet calculated that some two thousand million years must have elapsed since the primitive mountains emerged.

De Maillet's theory was as coherent as it was unorthodox. For the relative ages of his successive mountains, he depended on the principle of superposition. He could account for the facts that remains of marine life are not found in the oldest rocks but abound in the younger, even in strata of high mountains. Raised beaches and wave-cut benches rise like stairways along many parts of the Mediterranean shores.

What most exercised many of de Maillet's critics was his dismissal of the Holy Word as a source of factual information concerning the history of the Earth. The deluge may have been a catastrophic local event, de Maillet conceded, but it could not have been universal. He could forgive Noah for thinking otherwise. A question: if the flood had been universal, how then does one account for all the varicolored races of men living today?

Buffon's theory

Early in his scientific career, George-Louis Leclerc conceived the idea of producing an encyclopedia that would treat all natural phenomena systematically. The first three volumes of his *Histoire Naturelle, Générale et Particulière* appeared in 1749, a year after the first edition of *Telliamed* was published. At that time Leclerc was in the service of Louis XV as Keeper of the Jardin du Roi. For the previous ten years he had been cataloging the collections of a natural history museum which was part of the royal gardens. At the time of his death, 35 of the projected 50 volumes of the encyclopedia had been completed. Meanwhile the king had conferred on him the royal title Comte de Buffon.

Of all Buffon's voluminous writings, the most significant in a geological context was his *Epoques de la Nature* which appeared in 1778 as a supplement to the encyclopedia. (A definitive and analytical edition of this work was issued in 1962 under the editorship of Jacques Roger.) Here Buffon attempted a synthesis of Earth history from the beginning, speculated on the origin and development of life, and offered a numerical estimate for the Earth's age.

Buffon theorized that the planets of the solar system originated following the impact of a comet with the sun. The shock caused about 1/650 of the sun's mass to stream outward. That streamer of hot gases separated into globular masses to form the planets. In turn, the planets spun off satellites, fell back toward the sun and have continued to revolve around it ever since.

Thus during the first of Buffon's seven epochs the Earth was a globe of molten matter. During the second epoch the globe cooled and solidified from the surface inward. In the course of cooling, the outer crust developed cracks, blisters and wrinkles. The resulting irregular surface of high relief constituted the primitive mountains and plateaus.

During the third epoch the crust cooled to the point that water could condense and fall as rain. First pooled in depressions around the cooler polar regions, the primitive ocean ultimately covered all the Earth, except for local high prominences of the primordial crust.

With the appearance of water, life began. Buffon held that life is an inherent property of matter. He proposed that 'organic molecules' formed spontaneously by the action of heat on fluid substances after the surface had cooled to a critical temperature. Animals and plants aggregated from these molecules would first inhabit the polar regions, since there the critical temperature would first have been manifest. With continued cooling of the polar areas, the organisms adapted to that critical temperature had two options: either they could migrate to warmer environments equatorward, or they could stay at home and die.

Buffon conceived that once the universal ocean formed, it was soon populated with great numbers and kinds of shellfish, whose remains account for the abundance of limestone in the lower levels of the stratigraphic column. During the fourth epoch, some of these marine deposits became exposed as the waters began to withdraw from the continents. Prior to their emergence, however, these bottom sediments had been molded by currents into submarine mountains and valleys.

As the mechanism for diminution of the oceans and the corresponding emergence of the continents, Buffon proposed the episodic collapse of roofs above the enormous cavities beneath the megablisters of the primitive crust. With each collapse water rushed in to fill the void, sea-level sank, and the continents grew larger.

By the beginning of the fifth epoch, sea-level stood at about its present position. The seamountains and seavalleys had emerged and hardened to form the present topographic features of the lands. Over wide areas these transformed marine deposits retained their near-horizontal attitudes. But wherever the blisters had collapsed, the strata above had been deformed.

The major structural event of the sixth epoch was the opening of the Atlantic Ocean. Buffon was aware that remains of mammoths had been found both in Europe and in North America. He reasoned that a former land-bridge between these continents must somehow have subsided.

The Age of Man constitutes the seventh and present epoch. Buffon argued that man is distinguished from other organisms by the gift of reason. But simply because man is last on the scene, he cautioned, one must not conclude that he is the purpose of Creation. After all, life and thought, in modes we cannot visualize, must exist on other planets throughout the starry universe.

Leclerc's royal title derives from the Burgundian hamlet of Buffon, where he had inherited real estate from his father. There he constructed an iron foundry where, among other items, cannons for the French military were cast. The foundry and surrounding buildings stand today in a restored form. Most of the many tourists who come here are probably attracted by the international fame attached to the name of the builder. Scientists might appropriately pay a visit for a more particular reason, for it was here that the first geophysical experiments were conducted to the end of estimating the age of the Earth in years.

How long did it take for a molten globe the size of the Earth to cool to its present temperature? To address that question, Buffon had his workmen prepare ten balls of mixed iron and non-metallic substances graduated by half-inches to a maximum of 5 in (13 cm). These were then heated in succession to near the melting points. For each ball, the time required for cooling to air temperature in a nearby cave was then measured. As anticipated, the times required for cooling increased with increase in diameter of the balls. Observing the rate of increased times of cooling in the graduated set, he calculated that a sphere of molten matter the size of the Earth would require about 75 000 years to cool to its present condition. That was the figure Buffon published. His unpublished manuscripts that came to light in the following century contain much greater estimates of the Earth's age, ranging to nearly 3 000 000 years.

Werner's theory

From his father, manager of iron works at Wehrau in Saxony, Werner received early instruction in mineralogy. After taking courses in mining and metallurgy at the Bergakademie in Freiberg, he went to study law at the University of Leipzig. While there he wrote a treatise on the external characters of minerals which was published in 1774. The book brought him instant fame, and in the year of its publication he accepted an appointment to the faculty of the Bergakademie, aged 25. There he remained until the time of his death – a tenure of 42 years.

In 1786 Werner published his *Short Classification and Description of the Different Rocks*. In it he recognized four categories of rocks ordered according to age. Later he added a fifth unit, so that his stratigraphic column took the form shown below, beginning with what he considered to be the oldest in the series.

1. Urgebirge (primitive rocks): includes granite, gneiss, schist, prophyry and other kinds of crystalline rocks.
2. Uebergangsgebirge (transitional rocks): contains limestone, coarse sandstone and some rocks today classified as igneous. Fossils locally abundant.
3. Flötzgebirge (stratified rocks): includes all common types of sedimentary rocks and contains abundant fossils.
4. Aufgeschwemmte gebirge ('washed-up rocks'; alluvium): includes gravel and sands along streams in mountainous areas, as well as sand, bog deposits and peat in lowlands. Contains bones of quadrupeds as well as many fossils reworked from older units.
5. Vulkanische gesteine (volcanic rocks): includes lava and pyroclastic deposits.

It should be noted that Werner's transitional rocks are today regarded as Paleozoic in age. His stratified rocks are presently assigned to systems ranging from Permian through Tertiary.

In assembling this column, Werner derived much of his factual information from the work of earlier investigators. However, he went beyond bare facts to frame a Neptunist theory which soon became the ruling theory of the Earth, and which influenced geological thought until well into the 19th century.

Werner proposed that in an early stage of its history all the Earth was covered with an ocean overlying a solid floor characterized by high elevations and deep depressions. The ocean held in solution or suspension all the substances that would later form the planet's crust. Crystals of various minerals were first precipitated. These settled to form a blanket of granite draped over the elevations and depressions on the ocean's floor. Other kinds of crystalline rocks accumulated by the same process. Thus were the primitive rocks formed. Concurrently, sea-level was falling, perhaps due to decomposition of the water or its escape into outer space.

After further lowering of sea-level, solid particles began to settle to the bottom along with crystalline precipitates to form the transitional rocks. At some point in time islands began to emerge. Thus some small discontinuities in the sedimentary sequence necessarily developed, but Werner regarded these as minor. He considered the primitive and transitional deposits to be essentially universal in extent.

With continued diminution of the sea, the mostly clastic and richly fossiliferous flötz strata were deposited, though not universally. Deposition of scattered alluvial deposits followed as the sea approached its present level. Combustion of coal beds led to sporadic and localized volcanic activity, and to the formation of 'pseudovolcanic' layers below the surface. Together with the surficial ashes, cinders and lavas obviously of volcanic origin, these comprise Werner's volcanic rocks.

The widespread favor accorded Werner's theory was due not only to its simplicity but also because it could account for a variety of geological phenomena. For example, the cores of many mountain ranges, such as the Urals, are made of granite and associated crystalline rocks. Rock layers in the primitive and transitional units are commonly steeply inclined, as would be expected of chemical precipitates that can plaster themselves on the walls as well as the bottom of a container.

The theory was mainly gradualistic, but not entirely so. Werner visualized that the waters of the primitive ocean were very turbulent. Powerful currents occasionally scoured deep channels into the substrate to produce submarine mountains and valleys. As for the time required for development of the crust, Werner shied away from figures, content to observe that 'our Earth is a child of time and has been built up gradually'.

Werner's lectures on mining, mineralogy and related subjects attracted hundreds of bright students to Freiberg from all over Europe. Many of them afterward made substantial contributions to natural science. Alexander von Humboldt studied with Werner before undertaking his remarkable scientific expedition into South America. Unlike his star pupil, however, Werner wrote very little. Most of what we know about his ideas comes from the writings of his students. Thus the most detailed account, in English, of Werner's system of rock classification and theory of the Earth was written by Robert Jameson and published in Edinburgh in 1808.

3.2 HUTTON'S PLUTONIST THEORY

The Scottish Enlightenment

At the time Jameson's book appeared, Edinburgh was a center of activity in the intellectual movement historians have called the Scottish Enlightenment. During an interval beginning around mid-century and lasting into the 1820s, more notable contributions to literature, art, architecture, philosophy, economics, engineering and natural science came out of Scotland than could have been expected of a land relatively so thinly populated. Robert Burns, Sir Walter Scott, Robert Adam, Sir Henry Raeburn, David Hume, Adam Smith and James Watt are the names that come most readily to mind when the mind turns to the Scottish Enlightenment. However, there were many others who also sparked that movement, but who are little known outside specialist circles – among them James Hutton, who has been called the father of modern geology.

So far as geology is concerned, one of the more important developments during this Enlightenment was the establishment of the Royal Society of Edinburgh in 1783. As a charter member, Hutton felt obliged to contribute something to the reports which the fellowship had pledged to publish. In 1788 his *Theory of the Earth; or an investigation of the laws observable in the composition, dissolution, and restoration of land upon the globe* appeared in Volume 1 of the Society's *Transactions*. His reasoning in that essay ran as follows.

The Earth as a heat machine

The Earth functions as a machine whose purpose is to sustain life. However, in a habitable world decay is a necessary function. For animals require plant food; plants require soil; and soil forms by the decay of rocks. Decay is also manifest in the wearing down of the land by running water. 'Our fertile plains are formed from the ruins of the mountains.' Ultimately the streams carry soils and other earthy materials to the shores, where waves and currents transport that waste to the bottom of the ocean.

Thus in the course of time, which to Nature is endless, 'we may perceive an end to this beautiful machine'. With the wasting away of the continents by stream erosion, the time may come when the Earth is no longer capable of sustaining man and other terrestrial life.

Therefore, unless we can discover some mechanism by which this terrestrial machine is rescued from decay by natural processes we must arrive at one of two conclusions: 'the system of this Earth has either been intentionally imperfect, or has not been the work of infinite power and wisdom'.

Evidence for the workings of a restorative mechanism is not hard to find. Most of the world's land is made of rocks originally deposited as sediment at the bottom of the sea. The remains of marine life entombed in these rocks admit of no other conclusion.

The raising of marine sediments to form new continents – composed of the waste of former continents – has been accomplished by the expansive force of subterranean heat. To appreciate the power of this force, we only need to see it in action, for example during the eruption of Mt. Etna. There a column of molten rock has repeatedly been forced upward from a source below sea-level to discharge at great heights above.

Further testimony to the power of this thermal force is seen in the fracturing and contortion commonly displayed in strata uplifted from the bed of the sea. As originally deposited, these layered rocks would have been horizontal or nearly so.

Apparently this internal heat has operated globally throughout past times. Ancient lavas erupted from volcanos long extinct are widespread. In addition there are bodies formed from molten material that did not reach the surface but consolidated at depths below.

Thus the Earth may be considered as a heat machine. Volcanos serve as its safety-valves to ward off the devastating effects of earthquakes which otherwise would be attended by unnecessary elevation of the land.

Implications regarding the Earth's antiquity

The time required for the completion of just one of these cycles of decay and renovation must have been immense, judging by the slow rate at which the present continents are being destroyed by erosion. Considering the 'succession of worlds' preserved in the record of the rocks, Hutton concluded that 'we find no vestige of a beginning, – no prospect of an end' for this terrestrial machine. Cycles of dissolution and restoration of land upon the globe have recurred over lengthy spans of time in the past and may recur again and again in the future.

Hutton's academic and professional background

Hutton was a reluctant author. He was 62 years old when he published his theory, the main ideas of which appear to have already crystallized in his mind at the age of 34. At the University of Edinburgh he had elected a pre-medical course of study, and at the age of 23 had been awarded an MD degree from the University of Leiden. However he opted for a career as an agriculturalist, and so settled on a family farm near Jedburgh in the Southern Uplands of Scotland. After some 14 years of farming, he moved back to his native Edinburgh in 1763. A man of independent means, he moved quickly into the intellectual circles, freely discussed with his friends his ideas concerning the ancient history of the Earth, and after the establishment of the Royal Society felt honor-bound to make them public.

Kirwan's attack on Hutton's theory

The publication of Hutton's theory drew him more criticism than praise. Prominent among his detractors was Richard Kirwan (1733–1812), Irish chemist and mineralogist. Ardent Neptunist and fundamentalist Christian that he was, Kirwan's main objection to the Huttonian theory was that it is at variance with the Scriptural account of Creation. Hutton's observation of 'no vestige of a beginning' 'is contrary to reason, and the tenor of the Mosaic history, thus leading to an abyss from which human reason recoils'. In Kirwan's view, the implication that Hutton's system of successive worlds must have been eternal, smacked of atheism (Kirwan, 1793).

Turning to the record of the rocks, Kirwan insisted that primitive granite forms the basement of all continents. As for the constant washing of soils into the sea, that cannot be so because much sediment ends up in lakes. Many successions of layered rocks such as marble and slate contain no marine fossils, as should be the case had they been formed on the seafloor. As for the source of Hutton's subterranean heat, Kirwan asserted that there's not enough coal, sulfur or bituminous matter in the world to account for it.

Hutton's response to Kirwan

Kirwan's attack moved Hutton to write an expanded account of his theory, a lengthy work of two volumes published in 1795. Responding to Kirwan's slur about the 'abyss', Hutton countered with the observation that 'the abyss from which the man of science should recoil is that of ignorance'. A theory built on the premise that the Earth was designed to sustain life, he protested, can hardly be branded as atheistic. Viewed from that perspective, we can dispense with any appeal to a deluge designed to destroy life wholesale.

Hutton complained that Kirwan had misrepresented his ideas regarding the wasting of the continents, the occurrence of marine fossils in stratified rocks, the resources of subterranean heat and the age of the Earth. He had not claimed that soil is constantly washed into the sea, only that soil is necessarily, episodically and ultimately so transported. Nor had he proposed that the subterranean heat was fired by the burning of coal or other substances. He didn't propose to know the source of the heat, but recent volcanic eruptions and ancient lava flows prove that it is there now and has been for a long time. As for fossils, one would not expect to find them in every stone; over long periods of time many would have been destroyed by chemical or mechanical processes. Finally, with regard to his 'no vestige of a beginning', he didn't mean to imply that the world is eternal, only that 'in tracing back the natural operations which have succeeded each other, and mark to us the course of times past, we come to a period in which we cannot see any farther'. This is simply a confession of human limitations.

Hutton's fieldwork

The account of Hutton's theory was first read in 1785 at two meetings of the Royal Society. While awaiting its publication, he conducted field investigations around Scotland to gather factual support for his ideas. At Glen Tilt in the Grampian Mountains he found dikes of red granite cutting across the black country rock. His expressions of excitement were so remarkable that his guides thought he must have discovered a vein of silver or gold. Then on the Isle of Arran he found granite of Tertiary age that had intruded and uparched strata of Precambrian and younger age.

In sea cliffs at Siccar Point near St Abbs on the eastern coast of the Southern Uplands, strongly contorted strata of Silurian age are truncated and overlain by near-horizontal beds of Devonian sandstone (Figure 3.1). Hutton's good friend, John Playfair (1748–1819), accompanied him on an expedition to this locality in 1788. Playfair's recollection of his emotions on viewing that great unconformity clearly reflects Hutton's views on the immensity of past time and on the cyclic functioning of his terrestrial heat machine.

> On us who saw these phenomena for the first time, the impression made will not easily be forgotten... What clearer evidence could we have had of the different formation of these rocks, and of the long interval which separated their formation, had we actually seen them emerging from the bosom of the deep. We felt ourselves necessarily carried back to the time when the schistus on which we stood was yet at the bottom of the sea, and when the sandstone before us was only beginning to be deposited... An epocha still more remote presented itself, when even the most ancient of these rocks, instead of standing upright in vertical beds, lay in horizontal planes at the bottom of the sea, and was not yet disturbed by the immeasurable force which has burst asunder the solid pavement of the globe. Revolutions still more remote appeared in the distance of this extraordinary perspective. The mind seemed to grow giddy looking so far into the abyss of time...
>
> *Playfair, 1805*

Figure 3.1 Strata exposed in sea-cliffs at Siccar Point. Folded and fractured strata of Silurian age exposed in cliff are overlain by near-horizontal beds of Devonian age in foreground and along level surface on skyline. Photograph courtesy of Dr Arthur Richards.

3.3 THEORETICAL GEOLOGY TOWARDS THE END OF THE 18TH CENTURY

The obligatory catastrophism imposed on theories of the Earth cast in the context of the Mosaic chronology had almost disappeared toward the close of the century. Almost, but not quite. In 1799, Kirwan responded to Hutton's counter-attack in a book entitled *Geological Essays.* One of his aims was to demonstrate that his own theory, unlike Hutton's, was in accord with Old Testament writings. In his words, the 'account of the primeval state of the globe and of the principal catastrophes it anciently underwent, I am bold to say, Moses presents to us'. He then went on to offer a geological exegesis of the first nine verses of Genesis. The rest of the text was essentially a rehash of Werner's theory.

Most of the contemporary theories of Earth history, whether Neptunist or Plutonist, were more gradualist than catastrophist in tenor. De Maillet's theory was notably so, and Werner invoked nothing more exciting than increases in ocean turbulence and submarine slumping to account for discordances in the stratigraphic column. Buffon had indeed speculated that the planets had formed after a collision of a comet with the sun. Against that proposal, Hutton complained that Buffon had offered a theory 'not founded on any regular system, but upon an irregularity of nature, or an accident supposed to have happened to the sun'. We should not, he asserted, ascribe the production of the 'beautiful system of this world' to some catastrophe such as 'the error of a comet'.

Yet there were elements of catastrophism in Hutton's own mechanistic theory. He conceived that the continents normally waste away gradually as a result of day-to-day erosional events continued over indefinitely long spans of time. On the other hand, he granted the possibility that the continents may have sometimes been destroyed suddenly in the course of a single event. If the lands are uplifted from the sea by thermal expansion, the expanded matter must have become less dense than before. 'We may thus consider our land as placed upon pillars, which may break' and cause the continents to collapse back upon the seafloor. Furthermore, the uplift of the lands, no less than their foundering, Hutton thought might be catastrophic, considering the 'violent fracture and unlimited dislocation' of the uplifted strata.

The many Neptunists and the few Plutonists of the 18th century had advanced the cause of geology in several ways. Together they had offered alternatives to an obligatory catastrophism tied to a Mosaic chronology. The Neptunists had begun the laborious task of formulating a stratigraphy, a sequence of layered rocks arranged in chronologic order. Before the end of the century the systemic names Jurassic and Tertiary had appeared in the scientific literature.

Neptunist theories embodied the idea of a *direction* in the unfolding of the Earth's history that was lacking in the repetitious cycles of decay and renovation envisioned by Hutton. In their historical reconstructions, the Neptunists began by assuming an original state of the Earth, from which progressive changes to its present configuration were deduced. Getting rid of all the water posed an embarrassing problem. (See Laudan, 1987, for a perceptive account of Wernerian contributions to historical geology.)

Hutton reversed the order of investigation. He began with the present configuration, and sought to infer previous ones as formed by agencies manifestly now causing or capable of producing past changes. In the process, he offered a rational explanation for development of great angular unconformities such as the one at Siccar Point. Also he demonstrated that granite is more reasonably conceived as an 'unerupted lava' than as the product of chemical precipitation from a primeval universal ocean.

The debate between Wernerians and Huttonians sparked some lively exchanges in viewpoints at Edinburgh and elsewhere. Although this controversy persisted into the following century, it would soon be eclipsed by yet another – that one revolving around opposing views regarding the *tempo* of major changes in the Earth's configuration throughout geologic time.

4

Geology's heroic age

4.1 GEOLOGICAL ISMS OF THE EARLY 19TH CENTURY

The interval 1790–1820 has been called 'Geology's Heroic Age' (von Zittel, 1899). 'Age of Geoisms' would be an equally appropriate designation, especially for the first two decades of the 19th century. During those 20 years, Neptunism became institutionalized. Plutonism was popularized, and gained support from a nascent experimental geology. Catastrophism re-emerged, this time based largely on discontinuities in the history of life as interpreted from the record provided by fossils. The possibility that Noah's flood marked the most recent of many catastrophic episodes became the basis for an emerging diluvialism.

4.2 THE WERNERIAN SOCIETY

In 1800 Robert Jameson went to Freiberg to study with Werner. Upon his return to Edinburgh in 1804, he was appointed Professor of Natural History at the University of Edinburgh, a chair he occupied for the next half century. In 1808 he published his *Elements of Geognosy*, which as previously noted is the most complete contemporary account of the Neptunist doctrine in English (Jameson, 1976). That same year he organized the Wernerian Society of Natural History. When its first volume of *Memoirs* was published in 1811, the Society had enrolled three honorary, 43 resident, 79 non-resident, and 100 foreign members. The *Memoirs* were published sporadically until 1822 but around 1820 the contents shows that rigorous Neptunism had virtually been abandoned. Meanwhile Hutton's views were gaining support, thanks to Playfair's skill in presenting them to a larger public in a more readable form, and to Sir

James Hall's experiments in reproducing various kinds of igneous rocks in the laboratory.

4.3 PLAYFAIR'S *ILLUSTRATIONS*

Trained for the ministry, Playfair demonstrated such skills as a mathematician that he was appointed a professor of mathematics at the University of Edinburgh in 1785. After his wide-ranging intellectual interests beyond the mathematical became recognized, he moved to a chair in natural philosophy in 1805. During his years at Edinburgh he came to know Hutton as a friend, and so to understand how his friend had assembled the terrestrial heat machine. Following Hutton's death, he set about writing a treatise 'drawn up with a view of explaining Dr Hutton's *Theory of the Earth* in a manner more popular and perspicuous than is done in his own writings'. The result was Playfair's classic *Illustrations of the Huttonian Theory of the Earth*, published in 1802. Doubtless he was motivated to undertake this task because Hutton was no longer around to defend himself against the blasts published in Kirwan's *Geological Essays*.

It was largely through this book that scientists of the 19th century came to understand Hutton's dynamical geology. Playfair's prose is as lucid and well organized as Hutton's was opaque and repetitious. A sample of the Playfair style has already been given in the context of his visit to Siccar Point. Style aside, Playfair went beyond mere explication to add concepts of his own design. As an important example, he attacked the catastrophist interpretation of valleys, as in the following passage.

> Every river appears to consist of a main trunk, fed from a variety of branches, each running in a valley proportional to its size, and all of them together forming a system of valleys, communicating with one another, and having such a nice adjustment of their declivities, that none of them join the principal valley, either on too high or too low a level, a circumstance which would be infinitely improbable if each of these valleys were not the work of the stream that flows into it.
>
> If, indeed, a river consisted of a single stream without branches, running in a straight valley, it might be supposed that some great concussion, or some powerful torrent, had opened at once the channel by which its waters are conducted to the ocean; but when

the usual form of a river is considered, the trunk divided into many branches, which rise at a great distance from one another, and these again subdivided into an infinity of smaller ramifications, it becomes strongly impressed upon the mind that all these channels have been cut by the waters themselves; that they have been slowly dug up by the washing and erosion of the land; and that it is by the repeated touches of the same instrument that this curious assemblage of lines has been engraved so deeply on the surface of the globe.

In presenting Hutton's theory, Playfair cast aside almost all its original deistic framework and presented the scientific argument bare – warts and all. Some of the warts he removed, delicately. As an example, Hutton sometimes expressed himself as if he thought that all calcareous rocks are composed of animal remains. Playfair conceded that this proposition is more general than the facts warrant, and must certainly have been more general than Hutton intended. That Hutton's prose suggested otherwise was probably due to 'some incorrectness or ambiguity' of his language.

4.4 HALL'S EXPERIMENTS

Sir James Hall (1761–1832) is remembered as a founder of experimental geology. Though one of Hutton's close associates, he was reluctant to accept that part of Plutonist theory claiming that many different kinds of rock derive from matter originally molten.

Hall proceeded to put that claim to a test. To an iron foundry he brought samples of 'whinstone' (basalt and other kinds of rock Hutton considered to be of igneous origin), had them melted in crucibles and then allowed them to cool over measured lengths of time. He discovered that when cooling proceeded rapidly the melts turned into glass. On slow cooling, over a period of 12 hours or more, the molten glass crystallized to form substances much like the original rocks.

Hall read an account of his findings before the Royal Society in 1798, shortly after the death of Hutton (who had strongly opposed the procedure of seeking to discover Nature's grand design in the laboratory). In his paper of 1805 he emphasized the similarity of modern lava from Mt. Etna and elsewhere to the ancient Scottish whinstones, and concluded that 'so close a resemblance affords a very strong presumption in favour of Dr. Hutton's system'.

4.5 CUVIER'S CATASTROPHISM

Georges Cuvier (1769–1832) has been credited with establishing comparative anatomy and vertebrate paleontology as sciences. In 1812 he published a major work on the bones of fossil vertebrates which included an original theory of the Earth entitled *Discourse Préliminaire*. This preliminary account was promptly translated by Jameson and published under the old-fashioned title *Essay on the Theory of the Earth* (Cuvier, 1817). The book immediately became a popular mainstay of the catastrophist doctrine. Four editions were issued in Edinburgh. Several French editions were also published, but under Cuvier's more explicit title, *Discours sur les révolutions de la surface du globe*. Cuvier built his theory on what he considered to be empirical evidence, mainly provided by paleontology and structural geology.

In the area around Paris, sequences of Cretaceous, Tertiary and Quaternary sedimentary rocks crop out or are exposed in excavations. There Cuvier found that remains of quadrupeds belonging to species now living are confined to the most recent of these deposits. At lower levels in the stratigraphic column are found the bones of extinct species of elephant, hippopotamus, rhinoceros, and mastodon belonging to known genera or else very closely allied to them. Still lower in the column the bones belong to genera not extant, and remains of mammals are very rare near the base of the Tertiary. Clearly, then, during the later epochs of the Earth's history, whole populations of quadrupeds have been exterminated. The cause could not have been some nearby dislocation of the crust, for the strata of the Paris Basin remain essentially flat-lying. In this sequence, however, strata containing remains of marine organisms separate beds with bones of terrestrial vertebrates. Therefore recurrent flooding by the sea must have been an agency involved in the extinctions, Cuvier concluded.

Cuvier noted that whereas the strata flooring the lowest and flattest parts of the Earth are generally flat-lying, those in the foothills of the great mountain chains are inclined and otherwise deformed. Because these deformed strata underlie the horizontal beds of the lowlands, they must be the older of the two. Moreover these older sequences contain fossils unlike those in the beds above; and in most instances these fossils belong to extinct species. Therefore, he concluded, the rocks in the foothills provide evidence for ancient revolutionary events attending dislocations of the crust. These dislocations caused fluctuations in sea-level, attended by widespread destruction of many different kinds of organisms (Figures 4.1 and 4.2).

As possible additional evidence for the violence of his revolutionary events, Cuvier cited the presence of 'numerous and prodigiously large blocks of primitive substances scattered over the surface of the secondary strata and separated by deep valleys from the peaks or ridges whence these blocks must have been derived'. The reference here is to exotic boulders, such as those scattered over the southern foothills of the Jura Mountains. Their lithology clearly indicates sources far southward in the Alps. Playfair, and Hutton before him, had proposed that these 'erratics' had been transported from their outcrops by glaciers. Cuvier was uncertain as to the transporting agency, but suggested that the boulders may have rocketed to their present situations in the course of eruptions. As indicated by the following quotation, Cuvier's revolutions were sudden, violent and lethal.

> These repeated irruptions and retreats of the sea have neither been slow nor gradual; most of the catastrophes which have occasioned them have been sudden; and this is easily proved, especially with regard to the last of them, the traces of which are most conspicuous. In the northern regions it has left the carcases of some large quadrupeds which the ice had arrested, and which are preserved even to the present day with their skin, their hair, and their flesh. If they had not been frozen as soon as killed they must quickly have been decomposed by putrefaction. But this eternal frost could not have taken possession of the regions which these animals inhabited except by the same cause which destroyed them; this cause, therefore, must have been as sudden as its effect. The breakings to pieces and overturnings of the strata, which happened in former catastrophes, shew plainly enough that they were sudden and violent like the last; and the heaps of *debris* and rounded pebbles which are found in various places among the solid strata, demonstrate the vast force of the motions excited in the mass of waters by these overturnings. Life, therefore, has been often disturbed on this earth by terrible events – calamities which, at their commencement, have been perhaps moved and overturned to a great depth the entire outer crust of the globe, but

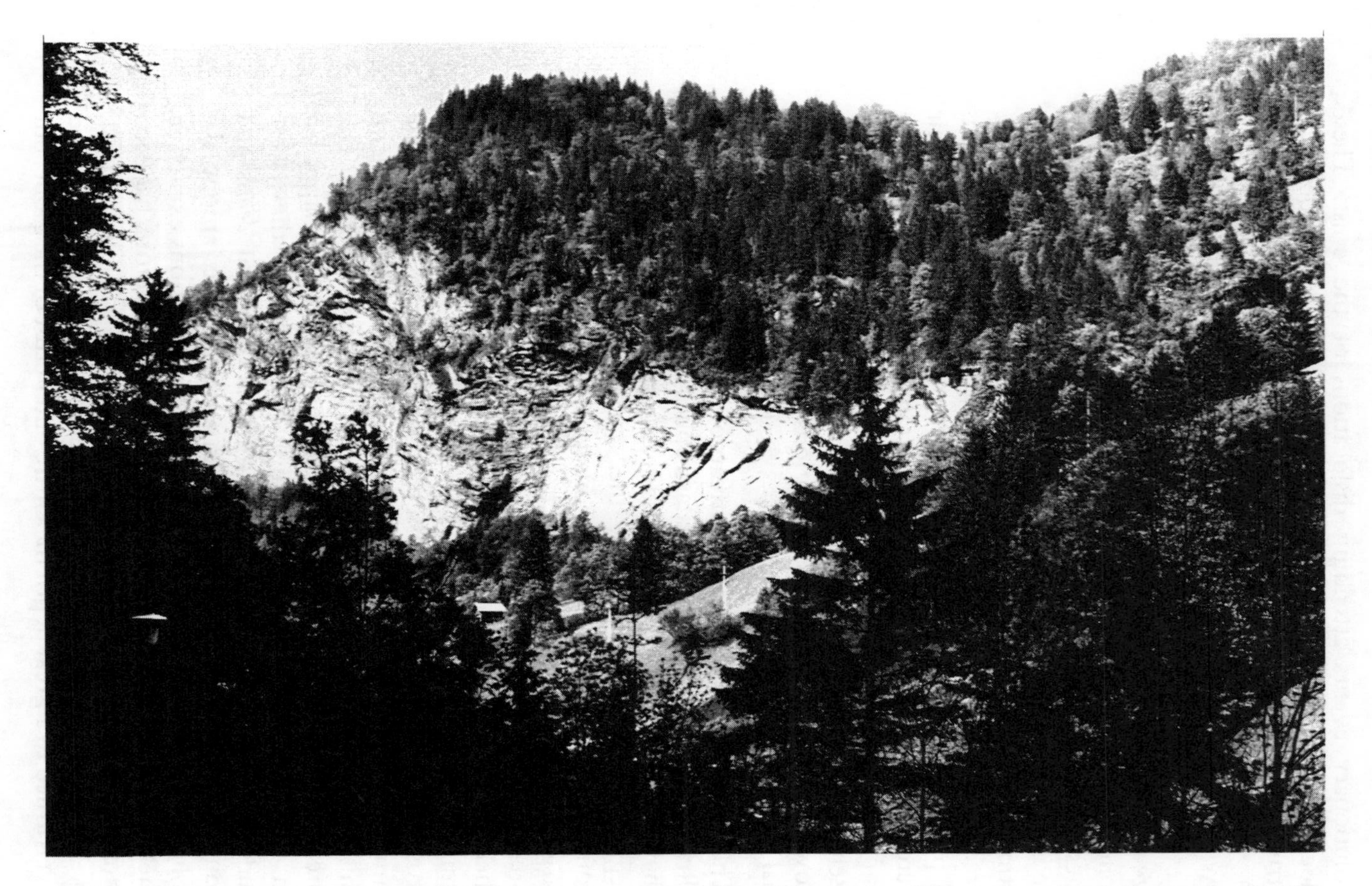

Figure 4.1 Strata bent into a recumbent fold, exposed in cliffs north of Lauterbrunnen, Switzerland. The presence of strongly deformed strata found in the Alps and elsewhere led Cuvier and others to infer that the Earth's ancient history has been punctuated by violent episodes of crustal dislocation.

Figure 4.2 Glacially transported granite boulder on southern flank of Jura Mountains north of Biel, Switzerland. Bedrock beneath the forest litter is limestone of Jurassic age.

> which, since these first commotions, have uniformly acted at a less depth and less generally. Numberless living beings have been the victims of these catastrophies; some have been destroyed by sudden inundations, others have been laid dry in consequence of the bottom of the seas being instantaneously elevated. Their races even have become extinct, and have left no memorial of them except some small fragment which the naturalist can scarcely recognize.
>
> *Jameson translation, 1817*

Cuvier believed that the last great revolution was witnessed by humans.

> . . . if there is any circumstance thoroughly established in geology, it is, that the crust of our globe has been subjected to a great and sudden revolution, the epoch of which cannot be dated much farther back than five or six thousand years ago; that this revolution had buried all the countries which were before inhabited by men and by the other animals that are now best known; that the same revolution had laid dry the bed of the last ocean, which now forms all the countries at present inhabited; that the small number of individuals of men and other animals that escaped from the effects of that great revolution, have since propagated and spread over the lands then newly laid dry; and consequently, that the human race has only resumed a progressive state of improvement since that epoch, by forming established societies, raising monuments, collecting facts, and constructing systems of science and learning.
>
> *Jameson translation, 1817*

Cuvier was willing to admit the veracity of the Hebraic tradition of the deluge. However, he insisted that Noah's flood was not a unique event, but only the most recent of many revolutions that have occurred over 'thousands of ages'.

One of Jameson's motivations to translate Cuvier's theory was to attack it as contrary to the teachings of Werner and Moses. In an appendix of notes running to some 150 pages, he corrected the Frenchman on three principal issues. The steep inclinations of strata commonly seen in high mountains are original, he claimed, and not the result of crustal dislocations. Erratic blocks found in the Juras and elsewhere were not erupted from their sources, but were transported

Figure 4.3 Boulders in terminal moraine along the Norwegian coast south-west of Larvik. Prior to acceptance of the hypothesis of continental glaciation, deposits such as these were commonly attributed to the work of catastrophic floods. Mounds of boulders on the skyline were constructed by the Vikings.

by the force of water, that is to say by the flood. As for Cuvier's 'thousands of ages', Jameson's emphatic response was 'Our continents are not of a more remote antiquity than has been assigned to them by the sacred historian in the book of Genesis, from the great era of the deluge' (Figure 4.3).

4.6 BUCKLAND'S DILUVIALISM

In 1813 William Buckland (1784–1856) was appointed Reader in Mineralogy at the University of Oxford. His title was later changed to Reader in Geology. His lectures were widely acclaimed, and many of his students acquired an abiding interest in geology.

The thesis of Buckland's inaugural lecture was that geological facts are in accord with Scriptural accounts of Creation and the flood. Unlike Kirwan, however, he did not subscribe to a literal reading of the Holy Word. Not days, but long periods of time might have elapsed between the successive acts of Creation, he allowed. As evidence for the flood he cited the presence of widespread blankets of 'gravel and loam' in situations where these deposits could not possibly have been laid down by streams. This diluvium he distinguished from ordinary alluvium found along rivers, and attributed its origin to universal flooding during Cuvier's latest revolution.

Buckland had formulated his theory before evidence accidentally turned up to give it what he considered proof. In 1821 quarrymen had exposed the opening to Kirkdale Cavern in north Yorkshire. When Buckland learned that numerous bones had been found there, he visited the cave and conducted a thorough examination of its fossils. Beneath a layer of mud along the floor he found remains of 23 species of animals. Skeletal parts of hyenas were by far the most numerous, but there were also identifiable bones of tigers, mastodons, rhinoceroses and hippopotamuses. Examination of tooth marks on bones led Buckland to conclude that the cavern had functioned as a den for hyenas in antediluvian times. The layer of mud overlying the bones he identified as a sediment left by the flood. Why should remains of antediluvian animals have been so concentrated and so well preserved in this cavern, Buckland wondered. His answer:

> ... the phenomena of this cave seem referable to a period immediately antecedent to the last inundation of the Earth, and in

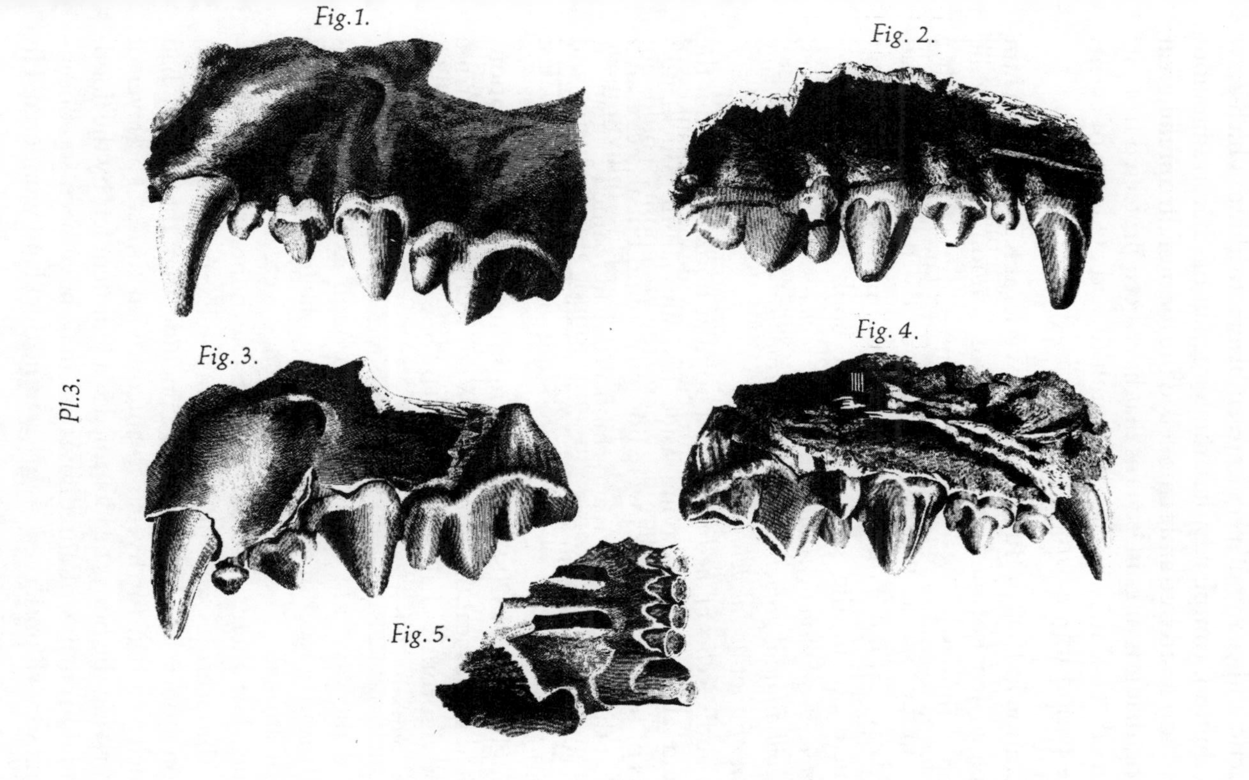

Figure 4.4 Reproduction of Plate 3 in Buckland's *Reliquiae Diluvianae*. Figures 1 and 2 show exterior and interior views of a modern hyena jaw. Figures 3 and 4 show analogous portions of fossil hyena jaws from Kirkdale Cavern. Figure 5 is a fragment of a jaw from Kirkdale showing incisor teeth much worn down.

> which the world was inhabited by land animals, almost all bearing a genetic and many a specific resemblance to those which now exist; but so completely has the violence of that tremendous convulsion destroyed and remodelled the form of the antediluvian surface, that it is only in caverns that have been protected from its ravages that we may hope to find undisturbed evidence of events in the period immediately preceding it. (Figure 4.4.)

Publication in 1822 of Buckland's discoveries at Kirkdale won him the Copley Medal of the Royal Society of London. Meanwhile he had been enthusiastically searching for antediluvial fossils in other caverns, in Europe as well as in England. By mid-1823 he had explored caves at more than 20 localities. A summary of his findings was issued late in the same year in a book entitled *Reliquiae Diluvianae; or observations on the organic remains contained in caves, fissures, and diluvial gravel, and on other geological phenomena attending the action of a universal deluge.*

Buckland's *Relics of the Flood* was a popular book throughout the 1820s and afterwards. Leaving aside his diluvialism, Buckland's accounts of fossil vertebrates were not only sound but also exciting. Proof that the hippopotamus and rhinoceros had roamed England at a not very distant time in the past was enough to fire anybody's imagination.

Throughout the 1820s catastrophism in its several formulations dominated geological thought in the West. Ironically, this doctrine was to be vigorously challenged in 1830 by one of Buckland's former students, namely, Charles Lyell.

5 Uniformitarians and catastrophists of the 19th century

5.1 LYELLIAN UNIFORMITARIANISM

Lyell's approach to historical geology

Charles Lyell's *Principles of Geology* was perhaps the most influential treatise on historical geology to come out of England during the 19th century. The first in his original edition of three volumes appeared in 1830; the other two followed in 1831 and 1833. Its long subtitle – 'being an attempt to explain the former changes of the Earth's surface by reference to causes now in operation' – foretells that the author will be following Hutton and Playfair in beginning with the present scene and then probing backward into the ancient past.

Like Hutton, Lyell would not spectulate on the origin of things. He would admit of no supernatural causes for changes in terrestrial configurations. He would not speculate on the age of the Earth as measured in years, but would provide evidence in support of Hutton's opinion that our planet must be almost inconceivably ancient.

Disagreements with Hutton

On the other hand, Lyell disagreed with Hutton on several important issues. While he accepted the Plutonist view that granite must be an intrusive igneous rock and not a primitive sediment, he rejected the proposition that sediments had been consolidated by subterranean heat prior to their elevation to form new lands. Nor would he accept

Hutton's suggestion that the elevation of continents might involve violent and paroxysmal convulsions.

More importantly, Lyell found Hutton's system deficient in that it underestimated the significance of fossils in the ordering of past events. That seems a little uncharitable, considering that Hutton was by far more interested in dynamics than in history. Furthermore, the principle of faunal sequence had not been clearly formulated during Hutton's lifetime.

Simply put, that principle states that in a sequence of fossiliferous formations, ordered according to their relative ages, different formations contain different kinds of fossils. In 1808 Cuvier and Brongnairt had established that fact in their investigations around Paris (Rudwick, 1976). In England, the principle was clearly enunciated by William Smith (1769–1839). Having completed his geological map of England, Wales and part of Scotland showing the areal distribution of 23 different rock units, Smith stated in his memoir describing those units that 'each stratum (i.e. formation) . . . has the same exterior qualities and the same extraneous or organized fossils throughout its course' (Smith, 1815). 'Extraneous fossils' and 'organized fossils' are now simply called fossils. Incidentally, Smith was a confirmed diluvialist and creationist, wary of materialistic theories of the Earth. As he put it in rhyme (Cox, 1942):

> Theories that have the Earth eroded,
> May all with safety be exploded,
> For of the Deluge we have data,
> Shells in plenty mark the strata.

Prejudices retarding the progress of geology

Upon reviewing the history of geology at the time of his writing, Lyell identified three sources of confusion. First, as dwellers on the land, we inhabit only about a fourth of the Earth's surface. The land is presently a theater of decay, where the continents are being gradually worn down by erosion. Little wonder then that some of the earlier theorists considered the world to be one ugly ruin. But we know that as the continents wear down, new strata build up on the floors of lakes and seas. Unfortunately, we are not in a position to observe this constructive phase of natural operations. Had mankind been born amphibious, we should have had a more balanced view of what goes on in the natural world.

Secondly, we are abysmally ignorant of processes in action at great depths below the lands, the hearth of igneous activity. Suppose, Lyell conjectured, we were in a position to observe the world from the bottom up rather than from the top down. Viewing the upward movement of molten rock toward the surface, invading and ingesting a 'roof' of fossiliferous strata, we might conclude that these strata are remnants of a substance that formerly encircled the entire globe. Assuredly our prejudices in matters geological have arisen from limitations in what is available to us for direct inspection.

Finally, preconceptions regarding the duration of past time have provided the greatest impediment to the advancement of geological knowledge. Lyell did not directly attack adherents to the biblical chronology. Instead he offered a fanciful analog to show that cramping into a narrow frame of time all former changes of the Earth's surface inevitably leads to a catastrophist scenario.

> How fatal every error as to the quantity of time must prove to the introduction of rational views concerning the state of things in former ages, may be conceived by supposing the annals of the civil and military transactions of a great nation to be perused under the impression that they occurred in a period of one hundred instead of two thousand years. Such a portion of history would immediately assume the air of a romance; the events would seem devoid of credibility, and inconsistent with the present course of human affairs. A crowd of incidents would follow each other in quick succession. Armies and fleets would appear to be assembled only to be destroyed, and cities built merely to fall in ruins. There would be the most violent transitions from foreign and intestine war to periods of profound peace, and the works effected during the years of disorder or tranquility would appear alike, superhuman in magnitude.

By the same token, Lyell argued, if we are parsimonious with time in our ordering of geological events then we must be prodigal with violence. If we deny any reasonable antiquity for the Earth, then the elevation of a great mountain range or an entire continent must be ascribed to some extraordinary or revolutionary cause. As an example, he cited episodic uplifts of the coast of Chile attending earthquakes of historical record. One of these quakes may raise the coast by an average height of about 3 ft (1 m) over a stretch of 100 mls (160 km). Two thousand such shocks might produce a mountain range 100 mls long and 6000 ft (1.8 km) high. If only one or two of these events occurred during a century, the situation would be the same as the Chileans have managed to tolerate. But if these shocks occurred during a single century, scarcely any animals or plants would survive and the surface of the country would be a confused heap of ruins.

Lyell's concept of the vast antiquity of the Earth was not simply cribbed from Hutton. He was an indefatigable traveler, and wherever he went he examined rocks and land-forms. For example, in 1828 he conducted a field study of Mt. Etna. There he found that this great volcanic mountain had been built up layer by layer of lava and ash, recording innumerable eruptions in prehistoric times. But he found nothing to indicate that the lava currents of remote periods were any greater than those witnessed by the Sicilians. Hence the mountain must have grown slowly over an immense span of time.

Lyell was tireless in offering gradualistic alternatives to catastrophist pronouncements. For example, mixtures of terrestrial and marine fossils in certain strata had sometimes been interpreted as evidence for the devastating revolutions envisioned by Cuvier. Lyell pointed out that great rivers in flood annually carry parts of terrestrial plants and bones of terrestrial animals into lagoons and shallows where they would be buried alongside skeletons of marine fish and shellfish.

The *Principles* immediately became a popular work, in no small measure due to the controversy it provoked. As Lyell continued to revise the text, it went through 12 editions, the last issued in 1875.

Lyell's choice of career

Had Lyell acceded to the wishes of his father, he would have earned his living by the practice of law. As a first step in that direction, he began studies at Exeter College, Oxford, in 1816. The following year he attended Buckland's lectures and went on occasional field excursions with his professor. Thereafter he acquired an unquenchable thirst for geological field-work. Though admitted to the bar in 1822, he opted for a career in science and about five financially lean years thereafter began writing his *Principles* (Wilson, 1972). His master-work is prominently noted in the inscription over his grave in Westminster Abbey:

Charles Lyell
Baronet F. R. S
Author of
The Principles of Geology
Born at Kinnordy, Forfarshire
November 14, 1797
Died in London
February 22, 1875
Throughout a long and laborious life
He sought the means of deciphering
The fragmentary records
of the Earth's history
In the patient investigation
of the present order of nature
Enlarging the boundaries of knowledge
And leaving on scientific thought
An enduring influence

To which most of us would say 'Amen'.

5.2 THE CHRISTENING OF UNIFORMITARIANISM AND CATASTROPHISM

In 1832, William Whewell (1794–1866) reviewed Volume 2 of the *Principles* at length, and in so doing first introduced the terms uniformitarians and catastrophists to differentiate the 'two sects' then represented in the geological community. In that review, and later in his *History of the Inductive Sciences from the Earliest to the Present Time*, he analysed the strengths and weaknesses of Lyell's theory (Whewell, 1872).

Lyell's basic assumption was that in order to build a reputable geological science, one must begin by assuming that the laws of nature are permanent and immutable – unchanging with the passage of time. Under that assumption, we may reason by analogy that the ancient causes of change in the world have been similar in kind to those now in progress. Thus we begin with the present scene and explore backward into the past (this is Gould's methodologic uniformitarianism). Lyell went beyond that to assume that past changes have been similar in degree to those operating in the present (Gould's substantive uniformitarianism; see Gould, 1965).

Whewell recognized that this substantive uniformity is not something given but instead is a hypothesis subject to testing. He agreed with Lyell that we are not arbitrarily to assume the existence of catastrophes. Even so 'the degree of uniformity and continuity with which terremotive forces have acted, must be collected, not from any gratuitous hypothesis, but from the facts of the case'. When Lyell 'considers it a merit in a course of geological speculation that it rejects any difference between the intensity of existing and of past causes, we conceive that he errs no less than those whom he censures'. That, Whewell observed, is not the temper in which science ought to be pursued. 'The effects must themselves teach us the nature and intensity of the causes which have operated; and we are in danger of error, if we seek for slow and shun violent agencies further than the facts naturally direct us, no less than if we were parsimonious of time and prodigal of violence'. The uniformitarians have arbitrarily selected the period in which we live as the standard for reconstructing the history of all earlier epochs. On what ground, Whewell enquired, shall we insist that man has been 'long enough an observer to obtain the average of forces which are changing through immeasurable time'?

5.3 SEDGWICK'S CRITICISM OF THE UNIFORMITARIAN DOCTRINE

In 1818 Adam Sedgwick (1785--1873) was appointed Woodwardian Professor of Geology at Trinity College, Cambridge University. Among other accomplishments, he is remembered for his naming of the Cambrian System, and jointly with Roderick Murchison the Devonian System as well. In an address before members of The Geological Society of London, delivered in 1831, he was sharply critical of Lyell on several important issues.

To begin, Sedgwick pointed out that if the Earth were originally molten, as suggested by the crystalline character of the oldest rocks, then the planet must have undergone a gradual refrigeration through time. Thus it seems unlikely that volcanic forces, to cite one example of a cause of change now in operation, could have acted at the same intensity throughout all time.

With Lyell, Sedgwick agreed that we must assume that natural laws are immutable.

> I believe that the law of gravitation, the laws of atomic affinity, and, in a word, all the primary modes of material action, are as immutable as the attributes of that Being from whose will they derive their only energy ... but (those) very powers themselves act under such endless modifications, sometimes combined together, and sometimes in conflict, that there follow from them results of indefinite complexity.

In other words, Lyell's theory 'confounds the immutable and primary laws of matter with the mutable results arising from their irregular combination'. When Lyell claims that 'no forces have been developed by this combination, of which we have not witnessed the results, he only proposes to limit the riches of the kingdom of nature by the poverty of our own knowledge'.

Sedgwick suggested that Lyell was thinking more like a lawyer than a historian when he proposed to see no progress in the history of life as revealed by fossils. Clearly, the successive appearances of higher forms of vertebrate life, including man, in the stratigraphic column indicate 'a progressive development of organic structure'. The laws of Nature, as we understand them, operating through secondary causes, cannot account for the recent appearance of man.

Discontinuities in the record of ancient life were attributed by Sedgwick to episodic sudden and violent revolutionary events separated by long periods of quiescence. As an example, he referred to the formation of the Jura Mountains. In his opinion, the broken and contorted strata there attest to violence; and after the building of these mountains there was an immediate change in many forms of life. As for the erratic blocks along the southern slopes of the Juras, these must have 'rolled off' to their present positions when the Alps were suddenly upthrust.

Reservations aside, Sedgwick conceded that nineteen twentieths of Lyell's work remained 'untouched' by his remarks.

5.4 LYELL'S RESPONSES TO HIS CRITICS

In his presidential address of 1850 to members of The Geological Society of London, Lyell looked back with satisfaction on developments during the twenty years elapsed since publication of the first volume of his *Principles*.

> I may, I think, affirm that the idea of comparing the modern agents of change with those of remote epochs, as not inferior in power and intensity, appears even to the most skeptical a far less visionary and extravagant hypothesis than when I first declared my belief in its truth.

The high degree of deformation, fracturing and metamorphism which had taken place in the Alps when that range was elevated during the Tertiary Period had been interpreted as the work of mechanical and volcanic forces of paroxysmal character. Lyell pointed out that this elevation was attended by deposition of rounded pebbles in accumulations up to 8000 ft (2.5 km) in thickness. Every pebble tells its own story: for all this rounding by stream action a long stretch of time would have to be inferred.

In the case of subsidence of the crust, no less than with its elevation, the process must have been gradual, Lyell insisted. As one line of evidence he pointed to the existence of vast thicknesses of marine sediments of Paleozoic age in Wales and elsewhere. The nature of the fossilized fauna and the physical characteristics of the rocks that enclose them both indicate slow subsidence in waters that remained shallow throughout the time of deposition.

In 1851 Lyell again defended his proposition 'that the ancient changes of the animate and inanimate world, of which we find memorials in the Earth's crust, may be similar both in kind and degree to those which are now in progress'. This time he boldly attacked the view held by many catastrophists that the fossil record provides evidence of a progressive development of life on Earth. In the case of extinct species, his arguments were twofold: the record provided by fossils is woefully incomplete; and many of these species were as 'perfect' as their living counterparts.

As for the apparent scarcity of mammals in Mesozoic rocks, Lyell proposed that this may be deceptive since few terrestrial deposits of that age have been found. Remains of birds are among the rarest of fossils, and so we should not despair of finding avian species in rocks as old as the Carboniferous. Mammals may also have lived in those ancient times, but left no remains.

Pursuing the 'perfection' argument, Lyell conceded that there have been four or five revolutions in plant populations since the Cretaceous. But he insisted that 'during these successive changes, there is no manifest elevation in the grade of organization, implying a progressive improvement in the floras which succeeded each other from the Eocene to our own epoch'.

Thus at the time Lyell delivered this address, he was not willing to believe that one species had descended from another by some evolutionary process. Faced with the problem of accounting for the appearance of a new species in the stratigraphic record, he simply confessed ignorance, but allowed for deistic as well as theistic possibilities.

> By the creation of a species, I simply mean the beginning of a new series of organic phenomena, such as are usually understood by the term 'species'. Whether such commencements be brought about by the direct intervention of the First Cause, or by some unknown Second Cause or Law appointed by the Author of Nature, is a point upon which I shall not venture to offer a suggestion.

It should be noted here that at the time Lyell delivered this address the standard geological systems were in place, with the sole exception of the Ordovician which was carved out of the Cambro–Silurian sequence as late as 1879. Much information had been gathered to indicate that each system contained distinctive fossils. Invertebrate animals in the oldest rocks were succeeded by vertebrates, with fish, amphibians, reptiles and mammals appearing successively in what appeared to be an orderly progression.

5.5 AGASSIZ AND THE DEMISE OF DILUVIALISM

As already noted, Playfair, and Hutton before him, had proposed that the erratic boulders of Switzerland were somehow related to transport by glaciers. Subsequent explorations by Swiss investigators, notably the engineer Ignace Venetz (1788–1859) and mining director Jean de Charpentier (1786–1855), had provided what they considered to be incontrovertible evidence that present Alpine glaciers had formerly been more extensive. That evidence included not only the far-traveled exotic boulders along the southern slopes of the Juras but also the striated pavements and moraines extending well beyond the ends of existing glaciers. Their ideas were not immediately accepted. One of the skeptics, Louis Agassiz (1807–1878), visited Charpentier and Venetz in 1836, viewed their evidence, returned to the Juras to reconsider that evidence in the light of the erratics there and, having seen the light, proposed in the year following a catastrophist hypothesis of continental glaciation worthy of a Cuvierian revolution.

Born in a Swiss village within sight of the Juras, Agassiz first rose to prominence as a scientist after he published a monograph on Brazilian fish in 1829, which he dedicated to Cuvier. Thereafter he began studies of fossil fish. He travelled to Paris in 1831, hoping to gain access to the paleontological collections there. Cuvier encouraged his work by turning over to him all his drawings and notes on the paleontology of fish. Agassiz's publications on that subject won him, at the age of 30, the Wollaston Medal of the Geological Society of London.

No wonder then that in 1837 members of the Helvetic Society, meeting at Neuchatel, expected their young president to deliver an opening address on fish. Instead, Agassiz talked about the ice-age.

At that time Agassiz held the catastrophist view that the history of life has been a record of alternate annihilation and creation of species. The ice-age, he now proposed, was brought on by the last eradication of life from the Earth. As for the ultimate cause he was uncertain, but as the collective body heat of the dead animals dissipated, the temperature of the Earth's surface dropped to freezing. All nature was enveloped in a shroud of ice. The Siberian mammoths were frozen. Glaciers formed over all North America and all of Europe southward to the Mediterranean.

Then, 'as a result of the greatest catastrophe which has ever modified the face of the earth' the Alps were thrust up. Huge blocks of rocks were propelled skyward, and landing on the uparched ice slid to great distances in all directions – hence the erratics. Following the reappearance of heat-generating organisms, the Earth warmed and the glaciers retreated to their present positions. Charpentier was embarrassed by this outburst, and von Humboldt earnestly advised Agassiz to give up speculations on glaciers and return to his research on fish.

In 1840 Agassiz published his classic *Etudes sur les glaciers*, a book he

---

Figure 5.1 Mid portion of Zermatt Glacier, Switzerland. Linear ridges of rocks (moraines) upon the glacier's surface are in process of slow transport down-valley toward the observer. Smooth rock surfaces to the right of the ice were abraded by the glacier when it was much larger. From Agassiz's *Untersuchungen über die Gletscher*, 1841; German edition of French original (1840).

dedicated to Venetz and Charpentier. In it he marshalled the factual evidence for the ice-age, but conceded that the comings and goings of glaciers may have been far more gradual than he had first supposed. Shortly after publication of the *Etudes* he toured England, Scotland and Ireland to find tell-tale evidence of continental glaciation. Buckland, who was sometimes a companion on these excursions, became as ardent a proponent of the glacial hypothesis as formerly he had been of diluvialism. It took about 30 years for the glacial hypothesis to become accepted in America and Europe (see Carozzi, 1984, for a succinct and well documented account of the history of glaciology) (Figure 5.1).

5.6 LYELL'S INFLUENCE ON DARWIN

If Charles Darwin (1809–1882) had bowed to the wishes of his father, he would have pursued a career as a physician. In 1825 Robert Waring Darwin, MD, sent his son Charles to the University of Edinburgh for medical training. While there, Charles attended Jameson's lectures on geology. As he later recalled, 'the sole effect they produced on me was the determination never as long as I lived to read a book on geology or in any way to study the science' (de Beer, 1974).

Having proved to his satisfaction that he was not cut out to become a medical doctor, Darwin enrolled at Christ's College, Cambridge in 1827 to train for Holy Orders – again on the advice of his father. There he became a close friend of Reverend Professor John Henslow, who had formerly offered courses in mineralogy and had recently moved to the Chair of Botany. Henslow encouraged Darwin to give geology a second chance. He arranged for the young friend to meet Sedgwick, who took him on a geological excursion through Wales in 1831. More importantly, it was Henslow who succeeded in arranging Darwin's appointment as naturalist to accompany Robert Fitzroy, Captain of H.M.S. Beagle on a voyage of exploration around the world. Early in the course of the expedition, which began on December 27, 1831, and lasted for almost five years, Darwin broke his vow never to read another book on geology.

> I had brought with me the first volume of Lyell's *Principles of Geology*, which I studied attentively; and this book was of the highest service to me in many ways. The very first place which I

> visited, namely St Jago in the Cape Verde Islands, showed me clearly the wonderful superiority of Lyell's manner of treating geology, compared with that of any author whose works I had with me or ever afterwards read.
>
> *de Beer, 1974*

The story of how Darwin, in the course of his travels abroad, conducted paleontological and zoological investigations which led him to consider that existing forms of life may have descended with modification from ancestral species has often been told and need not be repeated here.

In 1837, less than a year after the Beagle returned to England, Darwin set about gathering information relevant to the hypothesis of transmutation of species. That work continued intermittently for almost two decades, and in 1856 Darwin began writing a full account of his evolutionary theory.

> ... Lyell advised me to write out my views pretty fully, and I began at once to do so on a scale three or four times as extensive as that which was afterwards followed in my *Origin of Species*; yet it was only an abstract of the materials which I had collected, and I got through about half the work on this scale. But my plans were overthrown, for early in the summer of 1858 Mr Wallace, who was then in the Malay Archipelago, sent me an essay 'On the tendencies of varieties to depart indefinitely from the original type'; and this essay contained exactly the same theory as mine.
>
> *de Beer, 1974*

Darwin felt compelled to recommend publication of Wallace's essay. On the other hand, he hesitated to grant priority on ideas which already he had developed in greater detail. At the urging of Lyell and the distinguished botanist, Sir Joseph Hooker (1817–1911), he dashed off a summary of his theory. This was read on July 1, 1858, together with Wallace's account, before members of the Linnean Society. Soon afterward the two essays appeared in the Proceedings of the Society. Lyell and Hooker promptly urged Darwin to publish a book on the transmutation of species as quickly as possible. In a period of little more than a year he wrote the summary of 1859 which we know as the first edition of *The Origin of Species*.

Darwin's theory incorporated several ideas that were essential to Lyell's synthesis of historical geology. The first and most important had to do with the antiquity of the Earth. At one point in the text

Darwin advises the reader to close the book if he does not accept Lyell's views on the vastness of past periods of time. Naturalists who believed that species are immutable may be forgiven, he conceded, as long as they labored under the misconception that the Earth is only a few thousand years old.

Darwin also borrowed Lyell's gradualism – the idea that great changes occur in nature by summation of small changes over protracted periods of time. Natural selection, he proposed, acts by the preservation and accumulation of numerous small inherited modifications. Just as most geologists have abandoned the notion that a great valley has been excavated by some diluvial wave, so naturalists must abandon the idea that there have been great and sudden modifications in the structure of species.

Finally Darwin agreed with Lyell that the record of ancient life as revealed by fossils must be very incomplete. Given the fact that deposition of fossiliferous sediments has been episodic and not continuous, as indicated by the numerous unconformities in the stratigraphic column, one would not expect to find all the transitional forms linking existing species with their ancient predecessors.

What Darwin obviously did not borrow from his friend was Lyell's insistence on the immutability of species. However, in the fifth edition of *The Origin of Species*, Darwin triumphantly announced that Sir Charles had converted to evolutionism. Lyell was 72 years old at that time, something of a record for intellectual reversal.

5.7 THE KELVIN DISTURBANCE

William Thomson (1824–1907) is best known for his contributions to physics and engineering. He played a prominent role in the laying of the first Atlantic cable, and became wealthy following his invention of a receiver for the submarine telegraph. He was co-discoverer of the conservation law of energy, and was instrumental in devising the absolute temperature scale. At the age of 22 he was named Professor of Natural History at the University of Glasgow, and in the course of his long tenure there he published more than 600 scientific papers. In 1892, Queen Victoria made Thomson a peer of the realm, and he became Baron Kelvin of Largs (Kelvin from the name of a stream that flows near the campus of Glasgow University).

In 1852 Kelvin published an article entitled *On the universal tendency in nature to the dissipation of mechanical energy*. His argument was that while mechanical energy can neither be created nor annihilated through natural processes, it is constantly being transformed. In every change of energy from one form to another, he asserted, a portion of the original amount transforms to heat and is dissipated. Thus there can be no perpetual motion machine such as a self-winding clock, because the heat generated by friction between the moving parts constantly escapes from the system, and thus diminishes the store of energy available to swing the pendulum or otherwise move the hands.

Kelvin concluded that if the Earth functions as a heat machine, as Hutton had proposed, then it must be constantly losing energy as heat is conducted through the crust and dissipated through the atmosphere. Therefore:

> Within a finite period of time the earth must have been, and within a finite period of time to come the earth must again be, unfit for the habitation of man as at present constituted, unless operations have been, or are to be performed, which are impossible under the laws to which the known operations going on at present in the material world are subject.
>
> *Kelvin, 1852*

The sun is also spending its radiant energy at an enormous rate, and so must have been hotter than now in the geologic past, as it must assuredly grow cooler in times ahead.

Intermittently over a period of some 48 years Kelvin tried to calculate the age of the Earth. As one approach, he attempted to determine the length of time the sun has illuminated the Earth. His calculations were based on the assumption that the sun's heat had resulted from the collision of smaller masses. Once the sun had become incandescent, he proposed that it would behave like any other large mass of molten substance such as iron, silicon or sodium (Kelvin, 1871). Accordingly, he set 100 000 000 years as a reasonable limit for the time the sun has shone down upon the Earth.

Kelvin also addressed the question as to whether the heat from the sun, in its early stages of luminescence, was sufficient to sustain life on Earth. In this case he assumed that in the course of growth by collisions of its component parts, the sun's temperature would rise to a maximum and then decline after accretion ceased. On that basis, he calculated that the sun has been warm enough to support some form of animal and vegetable life for the past 20 or 25 million years only.

Kelvin's second line of approach to measuring the Earth's age was based on the outward flow of heat from the solid Earth. He assumed that the globe is cooling from an original molten state, and that most of the heat coming from the depths is residual from that state. It had long been established from investigations in mines that temperature increases with depth, though the rates of increase vary from place to place. Assuming that, on average, temperature increases by 1°F for every 50 ft (15 m) of descent, he attempted to calculate the time required for consolidation of the crust. However there were so many uncertainties involved in that exercise that he was only able to obtain what he considered as outside limits for the time past consolidation. In 1864 he concluded that crustal consolidation must have occurred not less than 20 000 000 years ago and not more than 400 000 000.

Subsequently, on the basis of new information relating to physical properties of rocks at high temperature, Kelvin greatly reduced the above upper limit. In an address before the Victoria Institute delivered in 1897 and published in 1899, he proposed that the time of consolidation was 'more than 20 and less than 40 million years ago, and probably much nearer to 20 than 40'.

Kelvin's figures for the age of the Earth were disturbing to the uniformitarians and evolutionists alike. Obligatory catastrophism, this time imposed by geophysics and not by Moses, cast serious doubts on Lyell's gradualism and Darwin's slow-working natural selection as the mechanism for organic evolution. (For an illuminating account of Kelvin's impact on geological thought, see Burchfield, 1975.)

Throughout all his investigations, Kelvin had insisted that the solar system must have evolved in accordance with the laws of matter. But he wisely provided a safety net for his geological theory in case it might fall under the weight of new evidence. Thus in his paper of 1871, referring to the history of the sun, he parenthetically observed: 'I do not say that there may not be laws which we have not discovered'.

5.8 DISCOVERY OF RADIOACTIVITY

In 1896, the year before Kelvin delivered his address before members of the Victoria Institute, the French physicist Henri Becquerel (1852–1908) discovered that compounds of uranium spontaneously emit energy possessing penetrating powers similar to those of X-rays. Soon afterward it was discovered that radioactive substances consistently give off heat. In 1904 Ernest Rutherford (1871–1937) proposed that radioactivity may be responsible for the heat emitted by the solid Earth.

> ... it does not appear improbable that the temperature gradient observed in the earth may be due to the heat liberated by the radioactive matter distributed throughout it. If this be the case, the present temperature gradient may have been sensibly constant for a long period of time, and Lord Kelvin's computation may only supply the minimum limit to the age of this planet. Thus the earth may have been at a temperature capable of supporting animal and vegetable life for a much longer time than estimated by Lord Kelvin from thermal data.... The discovery of the radioactive elements, which in their disintegration liberate enormous amounts of energy, thus increases the possible limit of the duration of life on this planet, and allows the time claimed by the geologist and biologist for the process of evolution.

Radiometric dating of rocks and minerals followed in the early decades of the 20th century. By 1937 Arthur Holmes was able to provide approximate dates for the beginning of each geological period back to the Cambrian (which he initially dated as beginning 470 000 000 years ago). Obligatory catastrophism, whether based on geophysics or metaphysics, would no longer be an issue in historical geology.

6

Meteorite craters

6.1 IMPACT AND EXPLOSION CRATERS

Historical geology of the 19th century was mainly directed toward inferring sequences of changes in the Earth's configuration produced by processes acting on the surface of the planet or seated in the depths below. The idea that bombardment by extraterrestrial objects could scar the landscape was not seriously considered until early in the 1900s, by those attending the debate about the origin of the much publicized Meteor Crater of Arizona.

That this large crater was formed by an explosion was apparent to most scientists who examined it. For a time, however, the view most favored held that the explosion was triggered by volcanic activity, and that by some odd chance vast numbers of meteorites had showered down on the very same place where a steam explosion had occurred. How the issue was at last settled in favor of an explosion caused by meteoritic impact is the subject of the following section.

Between 1921 and 1937 craters associated with meteorites had been reported from ten additional localities scattered over the world in both hemispheres. These are of two kinds.

Impact craters are formed wherever a meteorite has struck the ground and dug a hole in it. Most are less than 100 m across, and are bordered by low rims of ejected soil and rock fragments. Inside them, meteoritic fragments and dust are intermingled with soil or broken rock. Of 224 impact craters reported, some 200 were formed by the spectacular Sikhote–Aline shower that fell in the eastern Soviet Union in 1947. Most of the others are associated with larger explosion craters.

Explosion craters are formed by hypervelocity impacts of extraterrestrial bodies. They range in diameter from about 100 m to several kilometers. Around their rims bedrock is deformed and generally elevated relative to the center. Meteorite fragments are scattered over the surrounding area. Of examples known from 11 localities, 3 are solitary features and 12 are associated with impact craters. (For details and extensive bibliography, see McCall, 1977.)

For convenience, McCall has included the explosion of the Tungushka Meteor over Siberia in 1908 in his list of established meteorite craters. Because no craters resulted from that explosion, the unique Tungushka event will be treated separately (see p. 76).

6.2 THE METEOR CRATER OF ARIZONA

General features

The crater is a bowl-shaped depression in that part of the arid Colorado Plateau extending into north-eastern Arizona. It is about 180 m (600 ft) deep and 1.2 km (0.75 mls) in diameter. Its rim, capped with debris excavated from the crater, rises 30–60 m (100–200 ft) above the surrounding plain, which is underlain by nearly flat-lying strata of Permian and Triassic age (Figure 6.1).

Discovery of meteorites

In 1886 a company of shepherds pitched camp on the outer slopes of the crater and pastured their sheep nearby. They soon discovered that the area was littered with heavy metallic objects. News of the discovery attracted others to the site, among them mineralogist A. E. Foote who determined that the objects were iron meteorites. In 1891 Foote announced his findings at a meeting of scientists in Washington, D.C. Among those in attendance was Grove Karl Gilbert, considered by many to have been the most talented American geologist of his time.

Gilbert's hypothesis for origin

Gilbert immediately formulated the hypothesis that the meteorites and the crater were genetically related. He speculated that a shower

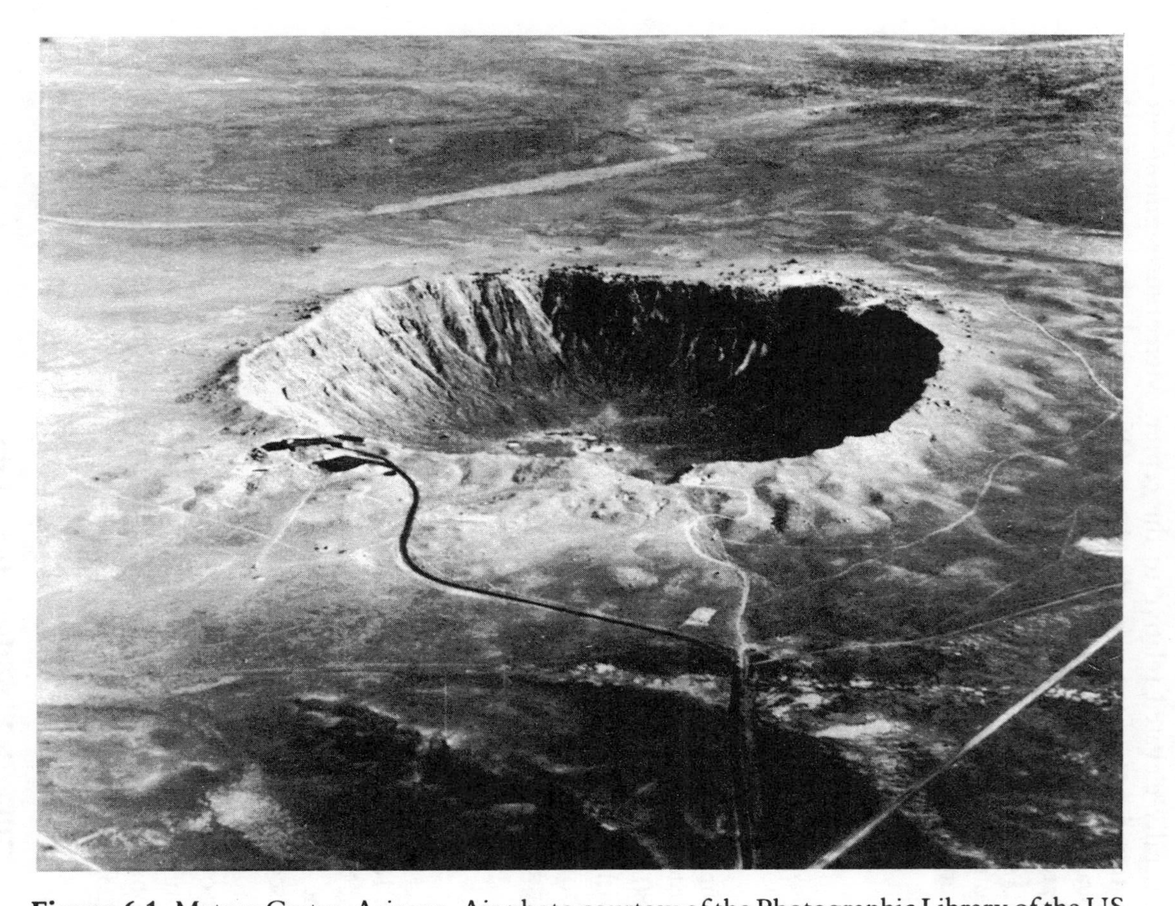

Figure 6.1 Meteor Crater, Arizona. Air photo courtesy of the Photographic Library of the US Geological Survey.

of meteorites included one larger than the rest which had produced the crater by the violence of its collision. Late in 1891 he set out for Arizona to conduct an investigation of his own. On the way, he formulated three critical tests of his impact hypothesis.

If the crater were formed by meteoritic impact, he reasoned, it should probably be elliptical in plan, since most meteorites would strike at an angle rather than fall vertically downward. If a large mass of iron were buried beneath the crater, it should deflect the magnetic needle of the compass. Furthermore, 'if a star entered the hole the hole was partly filled thereby, and the remaining hollow must be less in volume than the rim' (Gilbert, 1896).

Gilbert's impact hypohesis failed all three tests. The crater was not elliptical in plan. No magnetic anomaly could be found. Calculation of the volume of ejected rocks indicated that if all this debris were dumped back into the hole, it would fill it to the level of the surrounding plain.

Accordingly, Gilbert opted for the hypothesis that the crater formed by a steam explosion related to volcanic activity at depth. In support of that idea, he noted that the crater is in the midst of a great volcanic district, the nearest volcanic crater only ten miles (16 km) distant. However improbable, a shower of meteorites had fortuitously struck the same spot where a volcanically-induced explosion had opened a crater.

Barringer's investigations

Early in the 20th century Gilbert's abandoned impact hypothesis was revived by geologist and mining entrepreneur D. M. Barringer. Attracted by reports that some ten tons of meteoritic iron had been found around the crater, he organized a mining company, staked his claims to the depression and set about sinking shafts and drilling holes in search of a giant buried meteorite. It wasn't there.

Nevertheless, Barringer stuck to his impact hypothesis. Some of the sludge from the boreholes was nickeliferous, and specimens of oxidized nickel–iron had been found mixed with the ejecta along the rim. In his report of 1905, he argued that these mixtures indicated that meteoritic materials and ejected rocks must have been blown from the crater at the same time. (For an extended account of the debate about the origin of the Meteor Crater of Arizona, see Mark, 1987.)

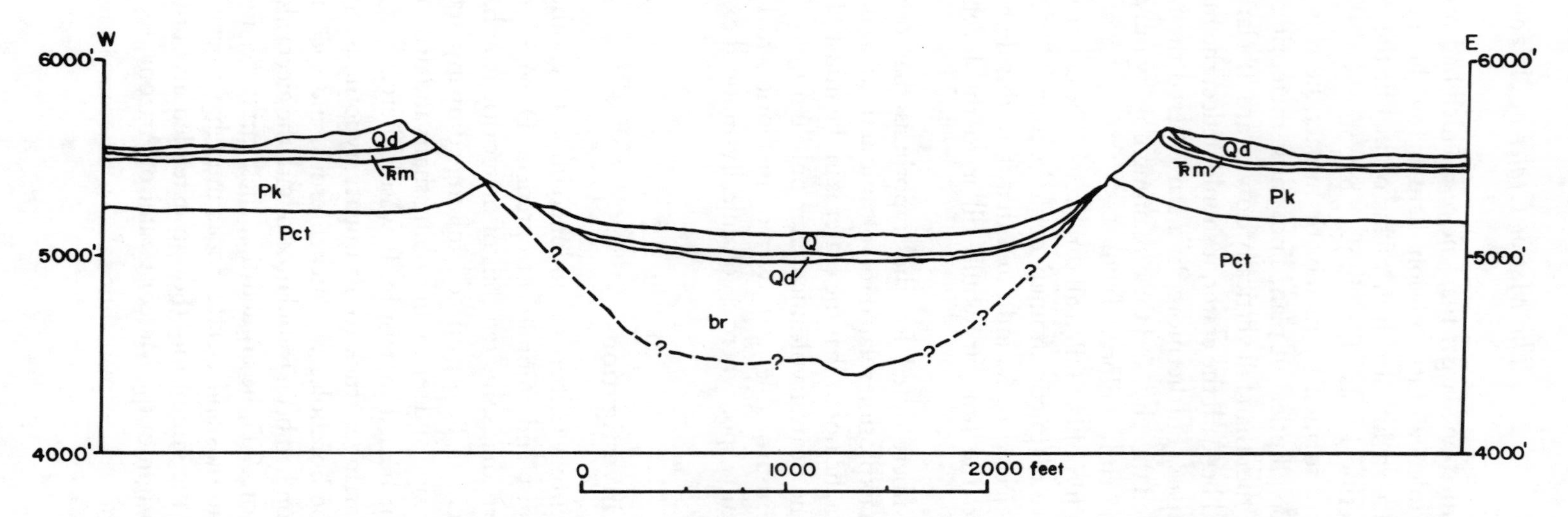

Figure 6.2 Cross-section of Meteor Crater, Arizona (after Shoemaker, 1960). Pct: Permian Coconino and Toroweap formations; Pk: Permian Kaibab Formation; Trm: Triassic Moenkopi Formation; br: breccia containing meteoritic materials; Qd: debris from Triassic and Permian formations; Q: Quarternary talus, alluvium and lake beds. Thin deposits of recent alluvium not shown.

Impact–explosive origin established

More recent investigations of local structural features and of the mineralogy and texture of ejected materials, combined with experiments on effects of high-velocity impacts, have established the Meteor Crater of Arizona as the prime example of a topographic feature formed by impact and explosion of an extraterrestrial body.

Both Gilbert and Barringer had reported that strata along the walls and in the rim of the crater dip radially away from the center. More recent investigations have shown that the angle of dip varies according to elevation. Low in the crater walls the strata are inclined gently outwards. Deformation increases upward towards the rim, where the beds are actually overturned along stretches which add up to about a third of the perimeter. These upturned and overturned layers are broken by small high-angle faults, most of which are parallel to regional sets of joints (Shoemaker, 1960) (Figure 6.2).

The sequence of strata exposed along the walls has been identified by Shoemaker as follows:

Strata	*Thickness* (ft)	(m)
TRIASSIC		
Moenkope Formation. Brownish fine-grained sandstone and siltstone.	30–50	9–15
········· Disconformity ·········		
PERMIAN		
Kaibab Limestone. Fossiliferous marine sandy dolomite, dolomitic limestone, and minor amounts of calcareous sandstone.	265–270	81–82
Toroweap Formation. White to brownish, calcareous, medium to coarse-grained sandstone.	9	2.75
Coconino Sandstone. White, fine-grained quartzose sandstone.	700–800	213–244

This bedrock stratigraphy is preserved in the debris surrounding the crater, but is inverted according to sequence, with fragments of the Moenkopi on the bottom and those from the Coconino on top. The blanket of debris consists almost entirely of angular fragments ranging from the microscopic to blocks more than 30 m across.

The floor of the crater is mantled by Quaternary talus and

alluvium, and toward the center by lake beds. Beneath these materials that accumulated after the crater formed is a layer of debris, a mixture of materials derived from all formations exposed in the crater combined with much oxidized meteoritic material. Shoemaker has interpreted this unit as having formed by fallback of debris thrown to a great height.

That the explosion which formed the crater generated enormously high temperatures and pressures is indicated by the metamorphic effects observed in quartz, notably in that of the Coconino. Locally the sandstone has been melted to form frothy masses of silica glass (lechatelierite), indicating temperatures around 1000°C or higher. In 1960 coesite, a high-pressure form of silica, was discovered as a subordinate component of silica glass. That substance had been synthesized seven years earlier but had never before been found outside the laboratory. Coesite is the stable phase of silica between 35 and 75 kilobars, pressures expected at depths of 60 to 100 km below the Earth's surface. Stishovite, another polymorph of silica that is stable only at even higher pressures, was also first found as a natural mineral at the Meteor Crater of Arizona (Chao, Shoemaker, and Madsen, 1960; Shoemaker and Kieffer, 1974).

Gilbert and Barringer had originally assumed that if the crater was formed by a meteorite, the main mass of that body must have lodged beneath the floor. During World War I, however, it was demonstrated that a bullet fired from a high-powered rifle and travelling at a velocity of 0.4 km/s would explode on hitting its target. That being the case, Wylie (1934) argued that large meteorites travelling at velocities 75 times that of the bullet must necessarily explode when they strike the ground.

Experimental evidence has confirmed Wylie's proposition. In 1963, Shoemaker, using a gas gun with hydrogen as the propellant, fired a small steel sphere into a specimen of Coconino sandstone. The bullet weighed 1.4019 g, and its velocity at impact was 4.27 km/s. In the explosion that followed a miniature crater 11–12 cm across and 2.45 cm deep was formed. The ejecta included strongly shocked quartz grains and small amounts of silica glass. Parts of the steel ball had melted to reform as minute spheres (Shoemaker, *et al.*, 1963).

Comparisons with craters formed by nuclear devices

As Shoemaker (1960) has pointed out, nearly all the major structural features found at the Meteor Crater of Arizona were reproduced in the alluvium of Yucca Flat, Nevada, attending the explosion of a nuclear device in 1955. There, a 1.2 kiloton device was detonated at a depth of 20 m below the surface. The resulting 'Teapot Ess' crater was about 90 m across and 30 m deep. As at the Meteor Crater of Arizona, the beds of alluvium in the rim are overturned. Debris around the rim roughly preserves the stratigraphic sequence in an inverted pattern. Glass is present in the debris. The floor and lower walls are underlain by breccia composed of fragments of alluvium mixed with glass.

Age

The age of the Meteor Crater of Arizona has been estimated at about 49 000 years (Shoemaker and Shoemaker, 1987).

6.3 OTHER SOLITARY EXPLOSION CRATERS

Wolf Creek Crater

Located in the East Kimberly District of Western Australia (lat. 19° 11′ S; long. 127° 48′ E), the crater is almost circular in plan. Its diameter ranges between 870 and 950 m. Prior to infilling by wind-blown sand and gypsum, it was probably 150–180 m deep.

Quartzite beds of probable Precambrian age dip radially outward from the crater, and in places are overturned and broken by radial faults. The rim is of fractured bedrock; any large accumulation of ejecta must have eroded away since the crater was excavated during Pliocene or early Pleistocene times.

Numerous 'shale balls', most measuring 5–25 cm across, have been collected from around the crater. These are made mostly of iron oxide, but some of them are also rich in nickel. They have been compared with the oxidized nickel–iron specimens Barringer found mingled with the ejecta at the Meteor Crater of Arizona. Unaltered meteoritic materials have not been found in the immediate vicinity of the crater. However, numerous small and heavily oxidized fragments of meteoritic iron are strewn over an elliptical area 3900 m to the south-west (McCall, 1965b; Taylor, 1965).

Boxhole Crater

This circular crater is named for nearby Boxhole Station in Central Australia (lat. 22° 54′ S; long. 135° 0′ E). It is excavated in an arid erosional plain developed upon metamorphic rocks of Precambrian age and mantelled with gravelly alluvium.

The depression measures 175 m across and is 16 m deep. Its rim rises about 3–5 m above the surrounding plain. Numerous fragments of meteoritic iron have been found in and around the crater. Small iron–shale balls have also been reported. Two puzzling features: no shattered blocks of bedrock appear along the crater walls and surroundings; and no impact glass has been found (Madigan, 1937).

6.4 IMPACT CRATERS

Sikhote–Aline craters

At mid-morning of February 12, 1947, many of the inhabitants of the Khabarovsk and Primorski districts in the Sikhote–Aline mountains geographic province of far eastern Siberia, witnessed a fire-ball streaking across the sky in a southerly direction. A trail of smoke and sparks followed in its wake. Shortly after it disappeared, loud explosions were heard. Pilots flying with the Soviet Air Force had observed the phenomenon, and they quickly pointed investigators to the place where meteorites had fallen.

Within an elliptical area of 1.6 km, more than 8000 nickel–iron meteorites were recovered (Krinov, 1966). Their aggregate weight measured about 23 tons. Evidently all of them were fragments of a single large body that disintegrated explosively several kilometers above ground level.

Fortunately, the field strewn with meteorites was 13 km removed from the nearest settlement. There were no casualties, although windows were shattered as far as 180 km from the strewn field, and earth tremors were reported as far away as 50 km.

The meteorites did not explode on impact. A few lodged in trees. Those that hit the ground formed some 200 holes and small craters. The largest crater reported by Krinov measured 26.5 m across and was 6 m deep. Craters were circular in plan and conical in cross-section, inner slopes slanting down at about 60°. In some cases the impacting meteorites had bounced out of the depressions they formed; otherwise they remained at the bottoms.

Haviland Crater

Beginning in 1885, numerous fragments of stony–iron meteorites were collected in increasing numbers from an area of several square miles on a farm near Haviland, Kansas (lat. 37° 37′ N; long. 99° 5′ W). In 1915 a shallow depression thought to be a buffalo-wallow proved on excavation to be the unfilled portion of an impact crater. Elliptical in plan, the principal diameters are 11 and 17 m. Inside the funnel-shaped depression, numerous meteorites were recovered, the largest weighing about 39 kg – nearly 86 pounds (Krinov, 1963).

Campo del Cielo craters

Early inhabitants of a semi-arid plain in what is now northern Argentina may have witnessed a meteorite shower even more spectacular than the one at Sikhote–Aline. When the first Spanish explorers entered this area, the natives told them a story about a large block of metal that had fallen from the sky. Translated into English, the Indian name for the area into which the metal had fallen is 'field of the sky'. In 1576 an expeditionary force located there a mass of iron with an estimated weight of 23 metric tons. Specimens were later identified as meteoritic iron containing a little more than 5% nickel.

Subsequent investigations have shown that at least nine craters, most with low raised rims, are aligned in a N 60° E direction over a stretch of 17.5 km (lat. 27° 35′–27° 40′ S; long. 61° 35′–61° 47′ W). They range in mean diameter between 20 and 100 m, and in depth from 0.5 to 5.5 m. Most are elliptical in plan, the largest with principal diameters of 91 × 115 m. A field strewn with meteorites, densest within a few kilometers of the line of craters, extends well beyond the first and last crater in the series. At least eight irons with masses of 100 to 4210 kg – 220 to 9262 pounds – have been reported. Two large specimens are displayed in the British Museum of Natural History and in the US National Museum.

Most of the craters appear to have been produced simply by impact. Cassidy *et al.* (1965) speculate that the craters and associated irons record the fragmentation of a parent meteoroid that entered the atmosphere at a flat trajectory. However, one crater (Hoyo Rubin de Celis), more nearly circular than the others, might be interpreted as having formed by explosion. Cores and trenches indicated that the crater was originally at least 14.8 m deep. Meteorite fragments and iron-rich shale were recovered from the fill. The burned stump of a tree was found standing upright near the rim; and farther away a pocket of charcoal was unearthed, its growth lines suggesting formation from a sizeable piece of wood lying horizontally. Cassidy and associates interpreted the charcoal as evidence of a forest fire lighted when the crater formed. The radiocarbon date for the charcoal is 5800 ± 200 years.

Dalgaranga Crater

This solitary crater is located in Western Australia (lat. 27° 43′ S; long. 117° 15′ E). Roughly circular in plan, it is 21 m across and 3.2 m deep, hence slightly smaller than the largest Sikhote–Aline impact crater (McCall, 1965a). It is excavated in granite and its overlying iron-rich crust. Several hundred specimens of meteoritic iron and stony-iron have been collected here. The age of the crater has been estimated at 25 000 years.

6.5 MIXED CLUSTERS OF IMPACT AND EXPLOSION CRATERS

Odessa craters

These are located on the arid plains of western Texas near the City of Odessa (lat. 31° 43′ N; long. 102° 25′ W). The principal crater is circular in plan, with a diameter of 168 m and a depth of 5 m. Excavations have established the original depth as about 27 m below the surface of the plain. More than 1500 irons, containing up to 7.55% nickel, have been collected from the surface about the rim. The crater is excavated mainly in flat-bedded limestone of Cretaceous age. These beds have been bent into an anticline that rings the crater (Krinov, 1963) (Figure 6.3).

Clearly the principal crater has been formed by explosion attending impact. Within a radius of 60 m from it are three small impact craters whose diameters range between 3 and 24 m. Nearly six tons of meteoritic irons have been recovered from excavations in these.

The age of the craters has been estimated at about 10 000 years (Mark, 1987).

Henbury Craters

Twelve craters are clustered in an arid plain over an area of 1.25 km² located about 50 km south of Alice Springs in Northwest Territory

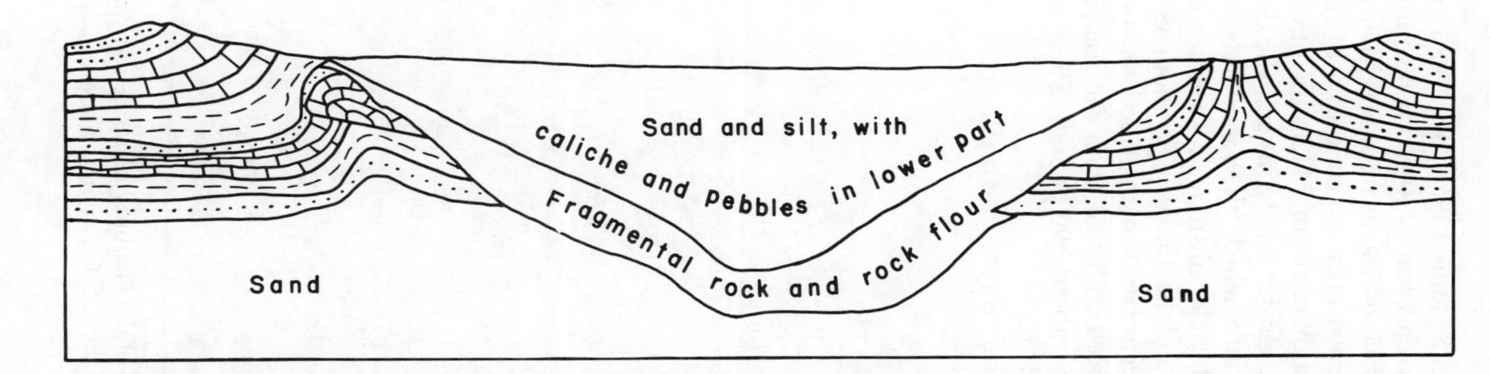

Figure 6.3 Cross-section of main Odessa Crater (after Krinov, 1963). Shows anticline in strata of Cretaceous age around periphery of crater. Meteorite fragments occur in basal unit of fill embedded in rock flour.

of Australia (lat. 24° 35′ S; long 133° 09′ E). The nine smaller ones range in diameter between 12 and 130 m. Of the four larger craters, two with diameters of 143 and 186 m are entire. Main Crater, with its oval rather than subcircular plan, is a composite of two overlapping depressions, one with a diameter of 242 m, the other 298 m across. The maximum rim height here is 12 m, and the depth 31 m. The Precambrian bedrock into which all craters have been excavated consists of interbedded shale, siltstone and sandstone. The strike is consistently east–west, and the dip averages about 35° S (Figure 6.4).

Several thousand specimens of meteoritic iron containing a little over 7% nickel have been collected from the area, the larger ones weighing up to 136 kg – about 300 pounds. Silica glass is scattered about but is not abundant.

The craters are especially interesting on three counts. The contrast

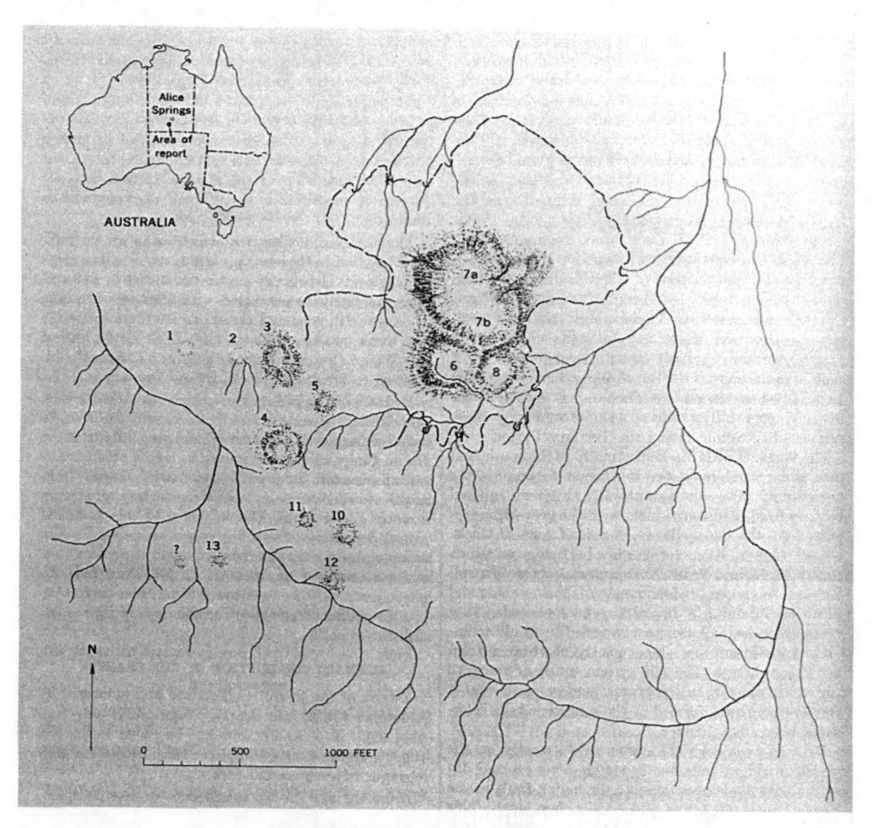

Figure 6.4 Outline map of the Henbury crater field (from Milton, 1972). Reproduced courtesy of the US Geological Survey.

here displayed between impact and explosion craters is striking. In two instances, ejecta from explosion craters formed 'rays' similar in plan to those extending from certain lunar craters. And, unlike the situations at the Meteor Crater of Arizona and Odessa where the target rocks lie flat, the strata here have fairly steep angles of dip so that the structures resulting from explosion tend to be asymmetric.

The smallest impact crater measures only 6 m across, hence less than a fourth the size of the largest Sikhote–Aline crater. Even so, it turned out to be a bonanza for collectors: nearly 205 kg or about 400 pounds of meteoritic iron was recovered following excavation to a depth of a little more than 2 m.

Ray-like trains of ejected sandstone blocks extend 60–70 m outward from the rims of two subcircular explosion craters. These craters are almost exactly the same size: in maximum diameter one measures 67 m across, the other 70 m.

Strata were displaced outward from all the explosion craters, but the effects of deformation around the rims differed according to whether the thrust was along or against the dip of the target rocks. This difference is clearly exhibited in one of the smaller craters which measures about 30 m across and has a maximum rim height of a little more than 1 m. On the updip northern wall the strata are apparently undisturbed. Around the opposite side the beds are folded and locally overturned.

Effects of far more powerful deformation appear around the Main Crater. Beds on the downdip side steepen to the vertical upward along the wall and then are overturned to form a flap of inverted strata. Locally the folded strata have been thrust outward over the pre-crater surface. Along the junction of the two craters that join to make the main one, imbricate thrust slices resemble the Alpine nappes on a small scale (Milton, 1972).

The age of the craters has been reported at 4200 ± 1900 years (Shoemaker, 1983).

Wabar craters

Two craters about a kilometer apart have been discovered in the central Rub al-Khali, the great sandy desert of Arabia (lat. 21° 29.5′ N; long. 50° 40′ E). The larger one is about 100 m in diameter and 12 m deep. It is clearly the result of explosive meteoritic impact. Aside from the evidence provided by iron meteorites scattered about it, there are vast quantities of vesicular silica glass concentrated around the rim. Coesite has been found in specimens of the glass, which also contains microscopic globules of nickeliferous iron. Glassy bombs have been collected from as far away as 40 m from the rim (Spencer, 1933).

The smaller crater is oval in plan, and 50×40 m in principal diameters. It is thought to represent the coalescence of two impact craters.

The age of the craters has been estimated at 6400 ± 2400 years (Shoemaker, 1983).

Kaalijaarv craters

On the island of Saarema in the Esthonian Socialist Republic of the Soviet Union, seven craters are scattered over a farmland area of $0.75\ km^2$ (lat. 58° 24′ N; long. 22° 40′ E). They are excavated in horizontally-bedded dolomitic limestone.

The main crater is circular in plan with a diameter of 110 m. Its rim rises 6–7 m above ground level, and encloses a pool about 16 m deep. Bedrock exposed along the rim dips radially away from the center at angles averaging around 60°. On the inside slopes of the rim the carbonate rock has been reduced to a powder.

The other craters range in diameter between 12 and 53 m. Excavators have turned up abundant particles of meteoritic dust and occasionally larger fragments of iron. Evidently the impacts that produced these depressions were not attended by explosions, for the strata in the walls and floors are crushed but not elevated (Krinov, 1963).

6.6 THE TUNGUSHKA METEOR

On the morning of June 30, 1908, a bright meteor passed northward over central Siberia. It was observed at localities as far as 710 km distant from its course. Eye witnesses have reported hearing rolls of thunder punctuated by three or four especially loud crashes. These booms were heard over an area of more than a million square kilometers – an area equivalent to that of France, the two Germanies, and Denmark combined.

Above a locality later identified as latitude 60° 55′ N and longitude 101° 57′ E, the fire-ball exploded at an altitude variously estimated to have been between 5 and 8.5 km above ground level. The resulting shockwaves in the atmosphere were recorded by barographs at English meterological stations a little more than five hours afterwards; they circled the Earth and were recorded at these same stations the second time around. Energy transmitted to the ground produced seismic waves recorded at an observatory in Irkutsk 893 km distant. On the first night after impact, and with lesser intensity on a few succeeding nights, the skies were unusually bright in western Siberia and all over Europe.

The area on the ground most severely affected by the shock is situated in a sparsely settled, heavily forested region some 200 km from the nearest village and located in the drainage of the Podkamennya Tungushka River. The first scientific expedition into this area was conducted in 1921. Other expeditions followed, and the 'Tungushka event' remains today a subject of study and speculation.

Members of the Russian expeditions discovered that trees had been blown down radially outward from a central area to distances between 40 and 50 km. Krinov (1966) estimated the area of forest destruction at about 2000 km^2. Dry timber had apparently been ignited by thermal radiation to distances up to 15 km from the central area.

No crater has been found, but Krinov has reported discovery of microscopic globules of magnetite and silica glass in soil around the devastated area. He proposed that these are residues from the explosion of a comet whose nucleus contained iron and silicates together with ice and gases such as methane, ammonia, and carbon dioxide. If so, this is the only known instance of a cometary impact on Earth in historic times.

Most more recent studies have tended to support Krinov's hypothesis. Glass (1969) conducted microprobe analyses of four of the silicate spherules and concluded that none has a composition similar to the silicate portion of any major meteorite group.

The nature of the exploding body, however, remains problematic. Ganapathy (1983) has proposed that the impactor may have been a stony meteorite. As evidence, he cites the fact that of eight submillimeter-sized spherules recovered from soil in the impact area, all contained traces of iridium ranging between 25 and 56 900 ppb, as well as traces of nickel and cobalt. On the assumption that if the debris from the explosion had entered the stratosphere its fall-out would have been global, he analysed a segment of an ice core taken at the South Pole. At depths between 10.15 and 11.07 m the iridium content increased over background by a factor of four. Dates assigned to these layers are in the range of 1912–1913 ± 4 and 1918 ± 4 years. Hence he proposed that the polar iridium anomaly records fall-out from the Tungushka event.

All investigators have agreed that whatever the nature of the bolide, its explosion was productive of an enormous amount of energy. Shoemaker (1983) has estimated that the kinetic energy released was equivalent to that produced by explosion of 10 to 30 megatons of TNT. That figure has been equated by Wasson (1985) to be about 500 times greater than energy released by one of the early atomic bombs.

Speculating on the chemical reactions that may have been triggered in the atmosphere, Turco *et al.* (1981) calculated that as much as 30 million tons of nitric oxide may have been generated in the stratosphere and mesosphere. The nitric oxide cloud, they proposed, may have depleted as much as 40% of the ozone layer in 1909, that reduction lingering over the next three years. Shoemaker has suggested that while the noctilucent clouds generated by the explosion follow a pattern to be expected during the Earth's encounter with the tail of a comet, the anomalous light may also have been due to chemical reactions involving nitrogen and oxygen.

Aside from massive destruction of trees, casualties to animals were remarkably small. Krinov estimated that about a thousand reindeer died in the area where the forest was leveled. Several persons living at distances as much as 60 km from impact center testified that they were knocked down and seared by a powerful hot wind.

6.7 THE RARITY OF METEORITE CRATERS

Explosion craters associated with meteorites are known from only 11 localities throughout the world. At an additional 16 localities, probable and possible impact explosion craters lacking meteorites have been cited (McCall, 1977). But if all those questionable examples are to be counted, meteorite craters would remain the rarest kind of topographic feature represented in the present landscape.

Like all small enclosed basins of whatever origin, meteorite craters are ephemeral. They fill with sediments, and their rims erode away. Not coincidentally, most established and probable examples are found in dry regions where stream erosion is relatively ineffective.

In the case of large impact–explosion features, however, tell-tale evidence as to origin may be preserved long after the craters have disappeared and the surrounding meteorite fragments have decomposed. That evidence should derive mainly from localized deformation and shattering of rocks attributable to explosive shock.

Certain anomalous circular structures similar to those produced by meteoritic impact have been found in rocks of various ages, some dating back to the Precambrian. Different investigators have given them different names, such as meteorite scars, astroblemes, cryptovolcanic structures, and cryptoexplosion structures. The debate concerning their origin has been no less intense than the one about the origin of the Meteor Crater of Arizona.

7

Cryptoexplosion structures

7.1 GENERAL FEATURES

Despite their wide diversity in age, size, and complexity, these structures are alike in two respects. They bear witness to violent localized explosions that shattered and otherwise altered rocks at the Earth's surface and to variable depths below; and no volcanic rocks of contemporary age have been found associated with them. They tend to be circular in plan, with diameters ranging between one and upwards of 100 km (McCall, 1979).

7.2 CONTROVERSY CONCERNING ORIGIN

The issue regarding origin, whether volcanic or meteoritic, was raised during the first decade of our century, at about the time Barringer was defending his unpopular views on the impact origin of the Meteor Crater of Arizona. In 1904, Werner tentatively proposed that the Ries Basin in southern Germany was formed by meteoritic impact. The following year, Branca and Fraas coined the term 'cryptovolcanic' for the smaller Steinheim Basin nearby.

In 1936, Bucher described six American structures which he proposed had been produced by the explosive 'release of gases under high tension, without the extrusion of any magmatic material, at points where there had previously been no volcanic activity'. In addition to citing evidence for the explosive origin of these near-circular structures, he pointed out that each is characterized by a central uplift ringed by a syncline. In one instance (the Wells Creek structure of Tennessee) the central uplift is encircled by peripheral folds decreasing in amplitude outward after the manner of damped waves.

That same year, Boon and Albritton proposed that Bucher's

cryptovolcanic structures could equally as well be interpreted as due to meteoritic impact. Among other matters they pointed to the bilateral structural symmetry, well displayed in several of the structures, as indicative of oblique impacts. A radial structural symmetry would be the expectable result of a gas explosion. The central uplift and encircling folds they attributed to elastic rebound of rocks compressed following the instant of explosion. As for Bucher's observation that the structures may have a causal relationship to certain broad domes and regional anticlinal swells, they pointed out that it would be hard to find a point in the interior of the United States that wouldn't bear a spatial relationship to major structures.

In 1963 Bucher and Dietz summarized arguments for and against the impact hypothesis in sequential papers published in *The American Journal of Science*. Bucher emphasized that Dietz's 'astroblemes' (star-wounds) are not randomly distributed as one would expect of impact–explosion structures. The European and American examples lie on large anticlinal flexures, while the Vredefort Dome of South Africa (cited as a possible meteorite scar by Boon and Albritton) 'is one of a whole string of basic and ultrabasic intrusives'. Coesite could have formed by explosive release of water vapor attending 'sudden crystallization of supercooled molten rock near the base of the crust'. Finally, traces of nickeliferous iron do not an astrobleme make, for that substance is common to basic and ultrabasic igneous rocks.

In his response, Dietz insisted that coesite and shatter cones validly indicate the intense shock best accounted for by the impact–explosion hypothesis. He could find no compelling evidence that the structures in question are genetically related to larger geological structures nearby.

In 1938, Boon and Albritton could identify only 14 structures considered to be what they called 'meteorite scars in ancient rocks'. With continued improvements in aerial photography and the increased use of air-photos in geological exploration combined with satellite imagery of the Earth's surface, the number of identified cryptoexplosion structures has increased dramatically. In 1982, Grieve cited 91 probable impact–explosion structures exhibiting shock metamorphism and scattered over the world. Additional examples continue to be added to the list, including one very large Precambrian structure in South Australia (Williams, 1986; Gostin *et al.*, 1986) (Figure 7.1).

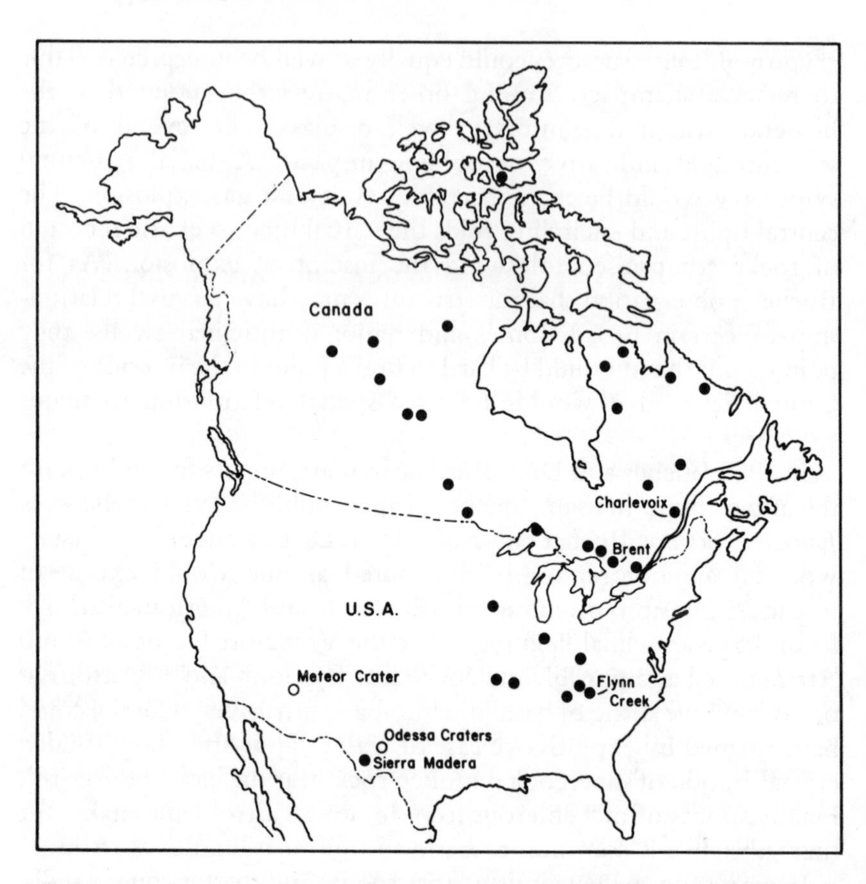

Figure 7.1 Map showing location of North American impact–explosion craters (circles) and cryptoexplosion structures (solid dots). Those named are mentioned in the text. After Sawatzky, 1977.

The first impact–explosion structure to be found in the ocean floor is located on the North Atlantic continental shelf 200 km south-east of Nova Scotia at a depth of 113 m (Jansa and Pe-Piper, 1987). The crater is at least 45 km in diameter. It is partly filled with breccia up to 850 m thick overlain by undisturbed marine sediments of Tertiary age. The breccia contains vesicular glassy particles. With regard to structure, the crater is marked by a central uplift 1.8 km high and 11.5 km across, ringed by a structural depression. Potassium–argon isotope dates range from 49.9 ± 2.1 my and 55.8 ± 0.9 my, hence in the age range of early to mid-Eocene. Melt rocks contain iridium, but in concentrations lower than would be expected from impact of an iron meteorite. The investigators speculate that the impactor was a stony meteorite or cometary nucleus with a diameter of 2–3 km.

7.3 A SAMPLER OF CRYPTOEXPLOSION STRUCTURES

Variation in size, age, structure and morphology

Five structures are here profiled. In diameter they range from 2.9 to around 160 km, in age from Precambrian to Miocene. One is a simple bowl-like depression; two others preserve 'fossil' craters, one with an upstanding central uplift, the other ringed by deformed strata and gigantic ejected blocks. For another, all traces of an original crater are gone, and the central uplift stands as a residual spike of breccia.

Brent Crater

Discovered in 1951 through examination of aerial photographs, the crater is located in Algonquin National Park of Ontario. It is 2.9 km in diameter and is excavated in crystalline metamorphic rocks of Precambrian age. Cores from boreholes indicate that the crater is underlain by upwards of 250 m of sedimentary rocks. Fossils recovered from that sequence have been identified as Middle Ordovician in age, suggesting that the crater is about 380 my old (Dence and Guy-Bray, 1972) (Figure 7.2).

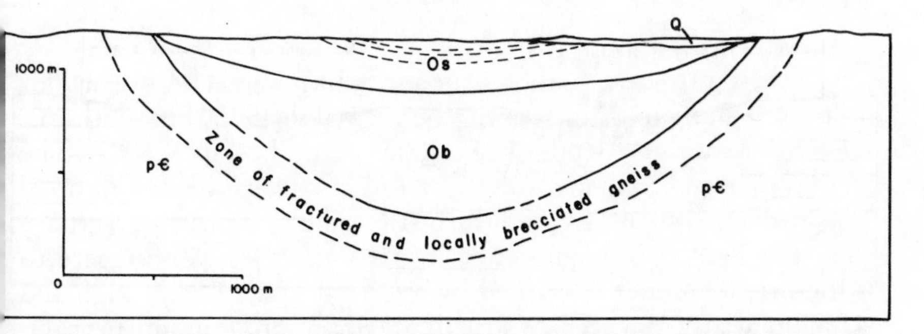

Figure 7.2 Cross-section of Brent Structure (after Dence and Guy-Bray, 1972). Ob: Ordovician breccia; Os: Ordovician sedimentary rocks; Q: Pleistocene glacial deposits.

The Ordovician strata are underlain by a lens of breccia about 610 m thick toward the center of the crater. Within the lower 35 m of that unit, fragments of shocked country rock are set in a matrix of once-molten matter subsequently recrystallized. Materials in this melt zone contain anomalously high traces of nickel, perhaps of meteoritic origin. Below the breccia the country rock is fractured and locally brecciated. Wall rocks exhibit similar effects of shock within distances as far as 400 m from the rim.

Dence and Guy-Bray cite Brent as a type example of impact–explosion craters of modest size formed in crystalline rocks. Tiny though it be in comparison with other Canadian cryptoexplosion structures, the magnitude of the brecciation here is impressive. The volume of breccia above the melt zone has been estimated at 2.46 km^3 (Dence, Grieve, and Robertson, 1977).

Steinheim Basin

Excavated in near-horizontal beds of Jurassic age, this circular basin is located in southern Germany (lat. 48° 02′ N; long. 10° 04′ E). It is about 3.4 km across and 90 m deep. In the center a hill 900 m in diameter rises 50 m above the otherwise flat floor. Miocene lake deposits are found on the flanks of the basin and beneath its floor. Results of drilling indicate that the depression was originally about 230 m deep, and the central hill some 150 m high (Figure 7.3).

Recent investigations have provided strong support for Rohleder's proposal of 1934 that the basin was formed by meteoritic

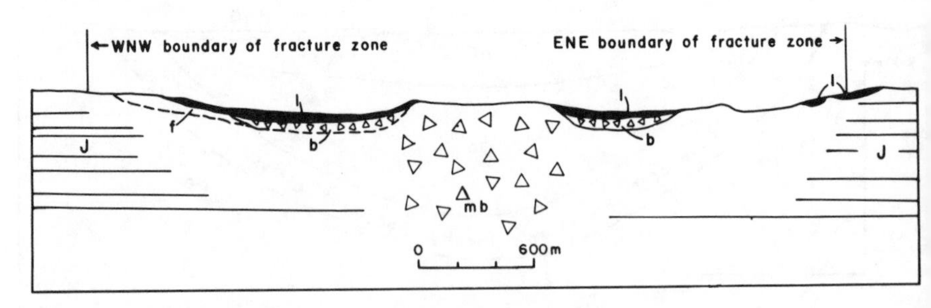

Figure 7.3 Cross-section of Steinheim Basin (after Reiff, 1977). J: essentially flat-lying strata of Jurassic age; mb: megabreccia of central uplift; b: fallback basin-breccia; f: zone of fault blocks; l: Miocene lake deposits.

impact and explosion (McCall, 1979). Strata around the flanks of the central hill dip radially outward at angles of 30° to 60°. Toward the center of the hill deformation is more intense. There the beds stand almost vertical within a megabreccia, some parts of which have been upthrust 250–380 m from their original stratigraphic positions. The Miocene lake beds overlie 20–70 m of breccia interpreted as fallback from an explosion. Many blocks in the breccia display well developed shatter cones; and sandstone fragments contain quartz grains with secondary planar features attributed to shock (Reiff, 1977).

After the basin formed, it was filled with lacustrine sediments and then was exhumed by stream erosion during the Quaternary. Lake

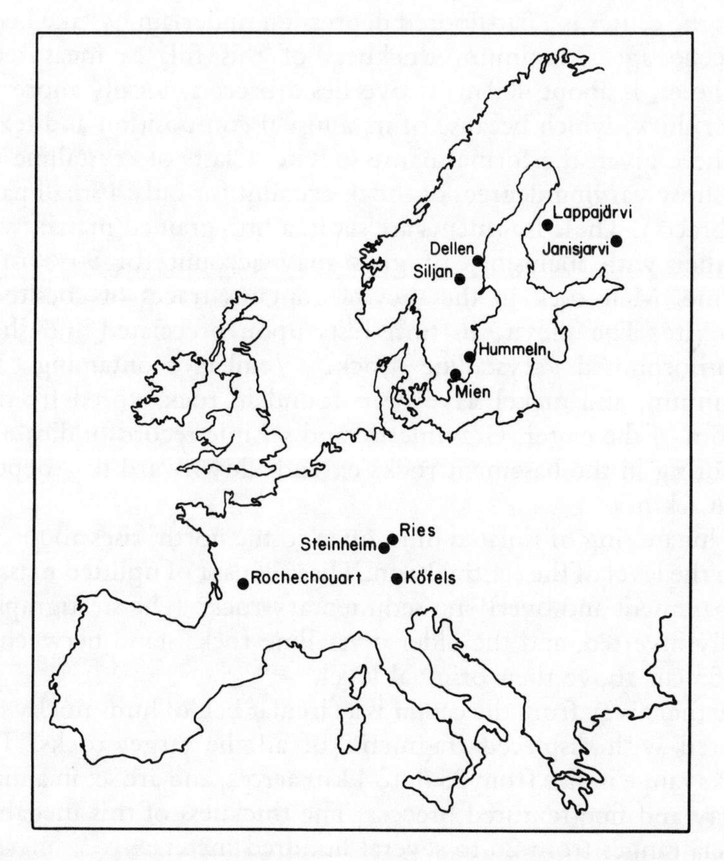

Figure 7.4 Map showing location of Reis, Steinheim, and other cryptoexplosion structures in Europe.

deposits of similar age found in the much larger Ries Basin 40 km to the ENE has led to the speculation that both these cryptoexplosion structures were formed at the same time, about 14.7 my ago (Figure 7.4).

Ries Basin

Perhaps the most thoroughly investigated of cryptoexplosion structures, the Ries is a circular depression about 24 km across located some 100 km north-west of Munich (lat. 48° 53′ N; long. 05′ E). In the area around the basin gently inclined sedimentary rocks ranging in age from Permian to Late Jurassic overlie a pre-Permian basement of granite and other crystalline rocks (Figure 7.5).

At the center is a flat-floored depression underlain by lake beds of Miocene age. Maximum thickness of this fill, as measured in boreholes, is about 400 m. It overlies a breccia, locally more than 400 m thick, which because of its unusual composition and texture was here given the formal name suevite. Clasts of crystalline rock that show varying degrees of shock account for only a small part of this breccia. These fragments are set in a fine-grained matrix which together with inclusions of glass may account for 90% of the volume. Melt rock in the suevite contains traces of coesite and stishovite. The suevite in turn rests upon brecciated and shock-metamorphosed crystalline rock. Veinlets containing iron, chromium, and nickel have been found in rocks cored from the bottom of the crater. Gravimetric and seismic records indicate that fracturing in the basement rocks extends downward to a depth of about 6 km.

A lunate ring of isolated hills, open to the north, rises about 50 m from the level of the central basin. These consist of uplifted masses of the basement and overlying sedimentary rocks. The stratigraphy is locally inverted, and the older crystalline rocks stand between 450 and 550 m above their original level.

Farther away from the center is a circular belt of hummocky relief covered with displaced fragments of all the target rocks. These blocks range in size from 25 m to 1 km across, and are set in a matrix of clay and fine-textured breccia. The thickness of this megablock breccia ranges from 20 to several hundred meters.

The rim of the Ries structure is marked by a circular system of downfaulted target rocks. Beyond that tectonic rim the strata lie undisturbed, but are overlain by patches of ejecta extending as far as

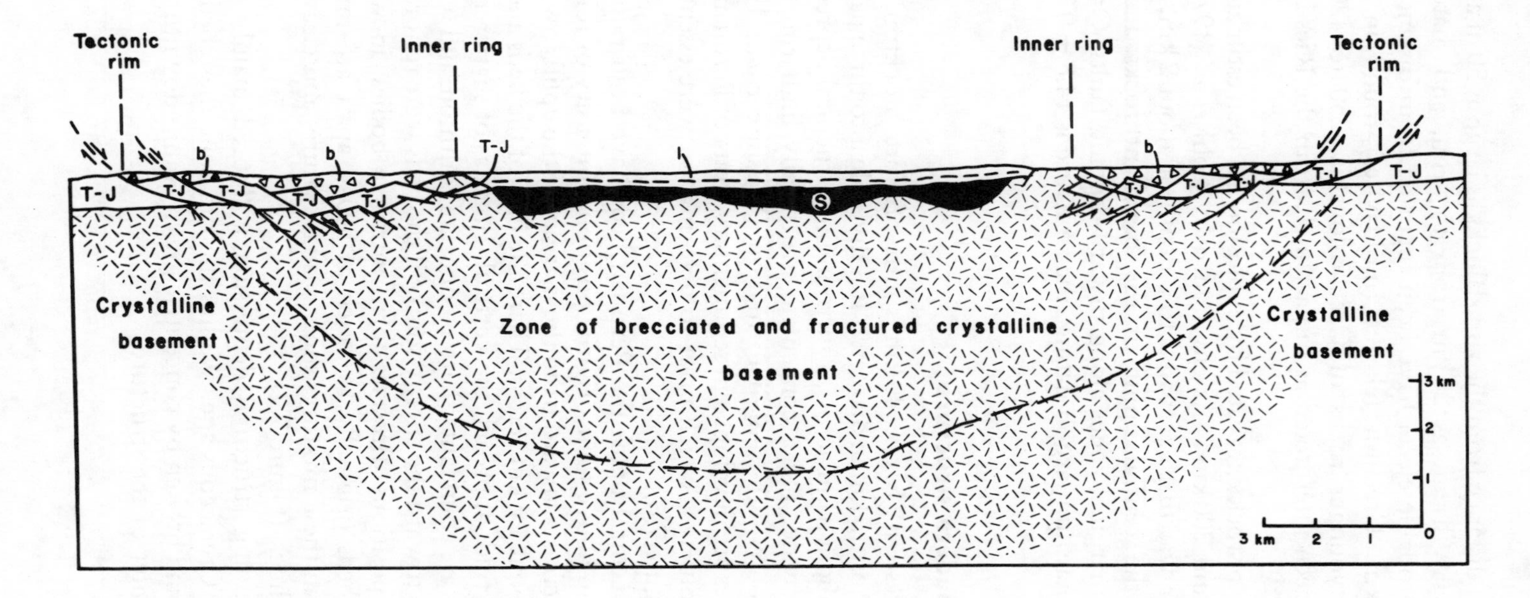

Figure 7.5 Diagrammatic section across Ries structure (after Pohl *et al.*, 1977). T–J: Triassic and Jurassic sedimentary rocks; S: suevite; b: breccia and ejected megablocks; 1: Miocene lake deposits.

40 km in radial distance from the rim. Blocks as large as 50 m across are found as far distant as 20 km from the rim, and materials identified as probable ejecta have been reported from distances as great as 350 km. Greenish tektites (moldavites) found in Late Miocene and younger rocks of Czechoslovakia, 250 to 420 km distant, have been interpreted as impact melt from the Ries event (Hörz, 1982).

The volume of rock excavated by explosion has been estimated as between 124 and 200 km^3. As envisioned by Pohl *et al.* (1977), a transient crater was first excavated to a depth of about 2 km, then modified by the central uplift of the basement rocks and by subsidence of marginal zones along inward-dipping faults. On the evidence of potassium–argon dating, the structure is 14.8 ± 0.7 my old.

Sierra Madera structure

Located in western Texas, 40 km south of Fort Stockton, this structure is almost 13 km in diameter. It is divisible into three parts: a central uplift about 8 km across, a surrounding structural depression 0.8 to 1.6 km wide, and a concentric structurally high rim zone marking the outer limits of deformation (Wilshire *et al.*, 1972). These structures are developed in sedimentary strata of Permian and Early Cretaceous age, which prior to deformation were essentially horizontal (Figure 7.6).

Within the central uplift the Permian beds have been raised as much as 1200 m above their original positions. Intensity of folding and faulting increases inward from the margins of the uplift toward a central zone about 1.6 km across. There the dips of the beds and the plunges of the folds approach the vertical. Breccias of shattered but unmixed rocks are abundant. Shatter-cones are common, and would point upwards toward the energy source if the beds were restored to their original positions. Breccias form intrusive bodies, in which fragments of beds originally as much as 500 m apart are mixed. Quartz grains in these mixtures show planar features and cleavages indicative of shock (Figure 7.7).

The surrounding structural depression is floored mainly with strata of Early Cretaceous age. Within the rim zone these beds are locally folded and broken by concentric gravity faults downthrown toward the center of the structure.

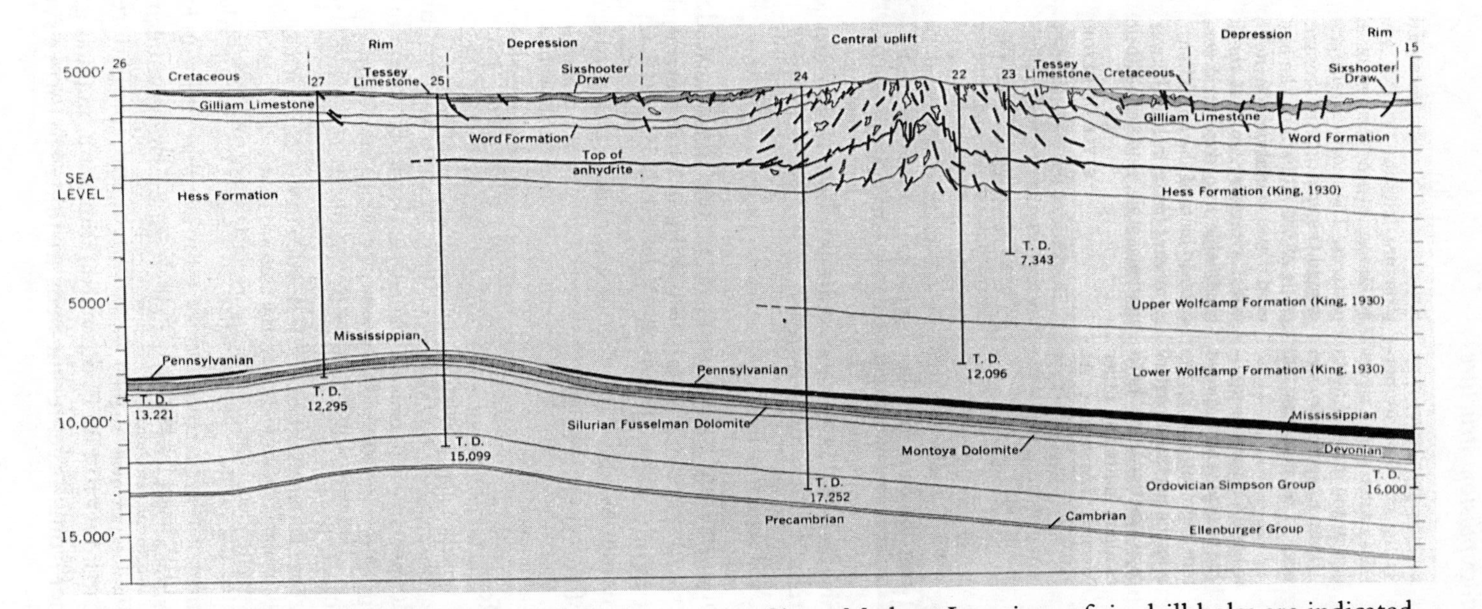

Figure 7.6 Generalized SW–NE structure section across Sierra Madera. Locations of six drill holes are indicated, and their total depths are given in ft. From Wilshire *et al.*, 1972. Reproduced courtesy of the US Geological Survey.

Figure 7.7 Shatter cones in Permian dolomitic limestone, Sierra Madera. Photograph reproduced with permission of Robert S. Dietz.

Exploratory drilling for petroleum around Sierra Madera has proved that the intense deformation exhibited in the surficial rocks disappears at depths around 1.6 km. That fact, conjoined with the evidence provided by the shatter-cones, makes a strong case for the hypothesis of meteoritic origin.

Vredefort Dome

This largest of circular structures that has been attributed to impact–explosion is located in South Africa to the south of Johannesburg. The deformed rocks are all of Precambrian age, and the structure is believed to have formed at some time during the Proterozoic Eon around 1970 ± 100 my ago (Grieve, 1982) (Figure 7.8).

At the center the Vredefort Granite crops out over a circular area 40 km in diameter. It is surrounded by younger Precambrian layered rocks measuring about 16 000 m thick. Around the granite margin these strata are for the most part overturned, and are shattered and metamorphosed over distances outward from the contact as great as 6.5 km. Farther from the center the beds are folded to form a syncline

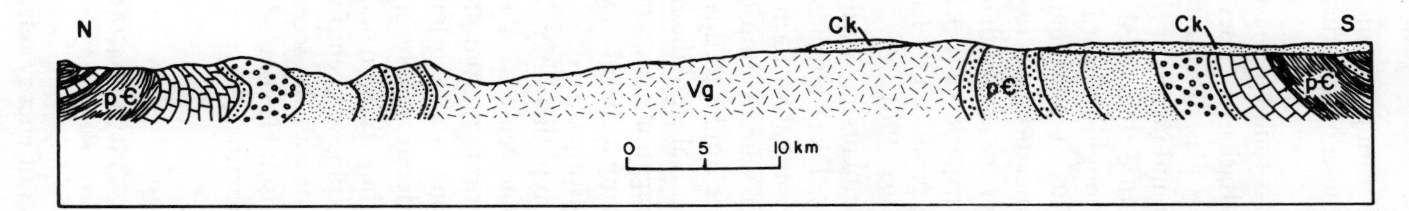

Figure 7.8 Idealized cross-section of the Vredefort structure. After Dietz, 1961. Vg: Vredefort granite; pC: Precambrian layered rocks; Ck: Carboniferous strata of the Karroo System.

that encircles the structure. This ring syncline measures about 60 km across, so that the overall diameter of the structure may be as much as 200 km (Dietz, 1961).

Evidence of intense shock is found in masses of pulverized rock, shattered or deformed crystals, shatter-cones and veins of materials thought to have originated as glass (pseudotachylite).

Shatter-cones have been found in almost all the different kinds of rocks in the Vredefort structure. If rotated back to their original positions they would point roughly toward the center.

The pseudotachylite consists of veins and stringers of jet-black microcrystalline materials measuring between 1 mm and 20 m across. It forms anastomosing networks both in the granite and in the encircling rim rocks. This rare species of rock has been interpreted as a melt (or 'shock impactite') produced by frictional heat attending passage of a shock wave (Dietz, 1961).

Overall, the structural elements at Vredefort display bilateral rather than radial symmetry. Dietz attributed this to an oblique impact by an asteroid. In quantitative terms, he suggested that a projectile 2.3 km in diameter with an impact velocity of 20 km/s could have formed a structure similar to this one. The energy thus produced would have been equivalent to the explosion of 1.4 million megatons of TNT. (One megaton equals the force exerted by a million tons of TNT.) For comparison, the explosion at the Meteor Crater of Arizona has been estimated as a 5–15 megaton event.

Opponents of the impact hypothesis have called attention to the fact that this structure is in the center of a larger structural basin, requiring that the hypothetical asteroid hit the bullseye. Furthermore, Vredefort is but one in a long line of major geological features, most of which are demonstrably of igneous origin.

According to Lilly (1981), the evolution of the Vredefort structure involved four stages: a first explosive event followed by a period of thermal metamorphism, and afterwards a second explosion followed by weak thermal metamorphism. He concluded that the structure is not that of an astrobleme, but rather the 'result of complex internal processes . . . not understood'.

7.4 EFFECTS OF EXPLOSIVE IMPACTS ON ORGANISMS

Considering the vast amounts of energy released by hypervelocity impacts of extraterrestrial bodies, these events must have been destructive to life on Earth. How destructive, outside the target area, remains a matter of speculation. In 1973 Nobel laureate H. C. Urey boldly proposed that collision with a comet was probably responsible for extinction of the dinosaurs and other organisms at the close of the Cretaceous period.

Urey assumed that a cometary mass of 1 000 000 000 000 000 000 g struck the Earth, travelling at a velocity of between 12.3 and 71.9 km/s (depending on whether the comet were stalking the planet or striking it head-on). In either case the explosion on impact would heat the atmosphere and ocean to levels intolerant to many forms of plant and animal life. Climatic changes would have been profound. Ocean water scattered over the lands would have caused widespread destruction of plants and animals. Dislocations of the crust would trigger giant earthquakes and volcanic eruptions.

Any dinosaurs and large air-breathing marine reptiles that survived the blast would be adversely affected by the incredibly high humidity of the atmosphere caused by vaporization of ocean water. Such moisture-laden air taken into their cool bodies would condense. Whether in or out of water the big reptiles would, in effect, drown from inhaling air, not water. The fact that alligators, primitive mammals, and birds survived the ordeal was mostly a matter of good luck. Not all areas were equally affected, and some animals were better able to cope with stress than others.

Urey's hypothesis seems not to have been taken very seriously by other members of the scientific community. That is, not until 1980, when discovery of a geochemical anomaly at the contact between Cretaceous and Tertiary strata led to formulation of the hypothesis that impact of an extraterrestrial body had been responsible for the mass extinction of species at the close of the Mesozoic Era. Which leads us to a consideration of sequential crises in the history of life.

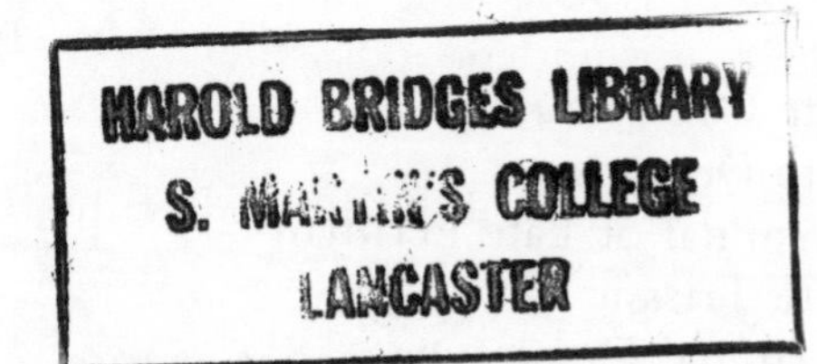

8

Mass extinctions

8.1 MAJOR EPISODES

The term 'mass extinction' has been applied to destructive episodes in the history of life, ranging in duration from less than a million up to about 15 million years, in the course of which unusually large numbers of species and higher taxa became extinct (Sepkoski, Jr, 1982). To quantify 'unusually large numbers', Raup (1982) proposed that more than about 10% of marine families or about 20% of marine genera would have to undergo complete extinction. Both figures, he suggested, would be higher for terrestrial organisms.

Kauffman (1986) recognizes three different concepts of mass extinction as expressed in current writings. Graded mass extinctions are usually attributed to earth-bound causes, such as major eustatic changes in sea-level, climatic changes including greenhouse effects, and intense volcanic episodes. Mass extinctions that have occurred as a sequence of discrete episodes over a stretch of 1–4 my he calls stepwise. For catastrophic mass extinctions, conceived to have occurred over periods of days, months, or a few years, extraterrestrial causes have usually been invoked.

Most paleontologists recognize the following five episodes of above normal extinctions (e.g. Raup, 1986b; Jablonski, 1986b) (see Figure 8.1 for geologic time scale).

Time	*Approximate my ago*
Late Ordovician	440
Late Devonian	365
Terminal or Late Permian	250
Late Triassic	215
Terminal Cretaceous	65

The numbers above are intended to mark terminations in extinction events, which may have begun long times before.

These major episodes are believed to be of different magnitudes. Sepkoski has divided them into two classes: major and intermediate. The Permian mass extinction stands alone as marking the most devastating collapse of the marine ecosystem. The remaining four, with terminations of 15–22% of marine families, are classified as intermediate. In addition, he identified five extinctions as 'minor'. That list includes events in the Cambrian, Early Jurassic (Toarcian), Late Cretaceous (Cenomanian), Late Eocene, and Pliocene.

Late Ordovician extinctions

All prominent groups of marine animals were affected. About a fifth of the one hundred or so marine invertebrate families became extinct. The trilobites lost 21 families, compared with 13 for nautiloid cephalopods, 12 for articulate brachiopods, and 10 for crinoids. Only two genera of graptolites survived (Sepkoski, Jr, 1982; Wilde *et al.*, 1986). Overall, some 57% of marine genera disappeared, possibly in the course of several million years (Sepkoski, Jr, 1986a).

Late Devonian extinctions

These occur at the transition between the Frasnian and Famennian stages. Cosmopolitan faunas of shallow seas covering much of the continents around the world were affected. The Frasnian was a time when widespread carbonate banks were formed by algae and corals, associated with a rich invertebrate fauna. After the extinctions, reefs virtually disappeared and remained rare until the Early Mississippian. Rugose and tabulate corals lost 25 families, articulate brachiopods 17, and cephalopods 14. Crinoids, the armor-plated placoderm fishes, and ostracods were also severely reduced in diversity, each of these groups losing 10 or more families. Approximately 86% of brachiopod genera failed to survive (McLaren, 1982; Sepkoski, Jr, 1982; McGhee, Jr, 1982). On the evidence of studies in south-central Asia, Farsan (1986) concluded that this extinction was not a single event, but was an accumulation of many successive episodes of extinction spread over a period of 6–8 my.

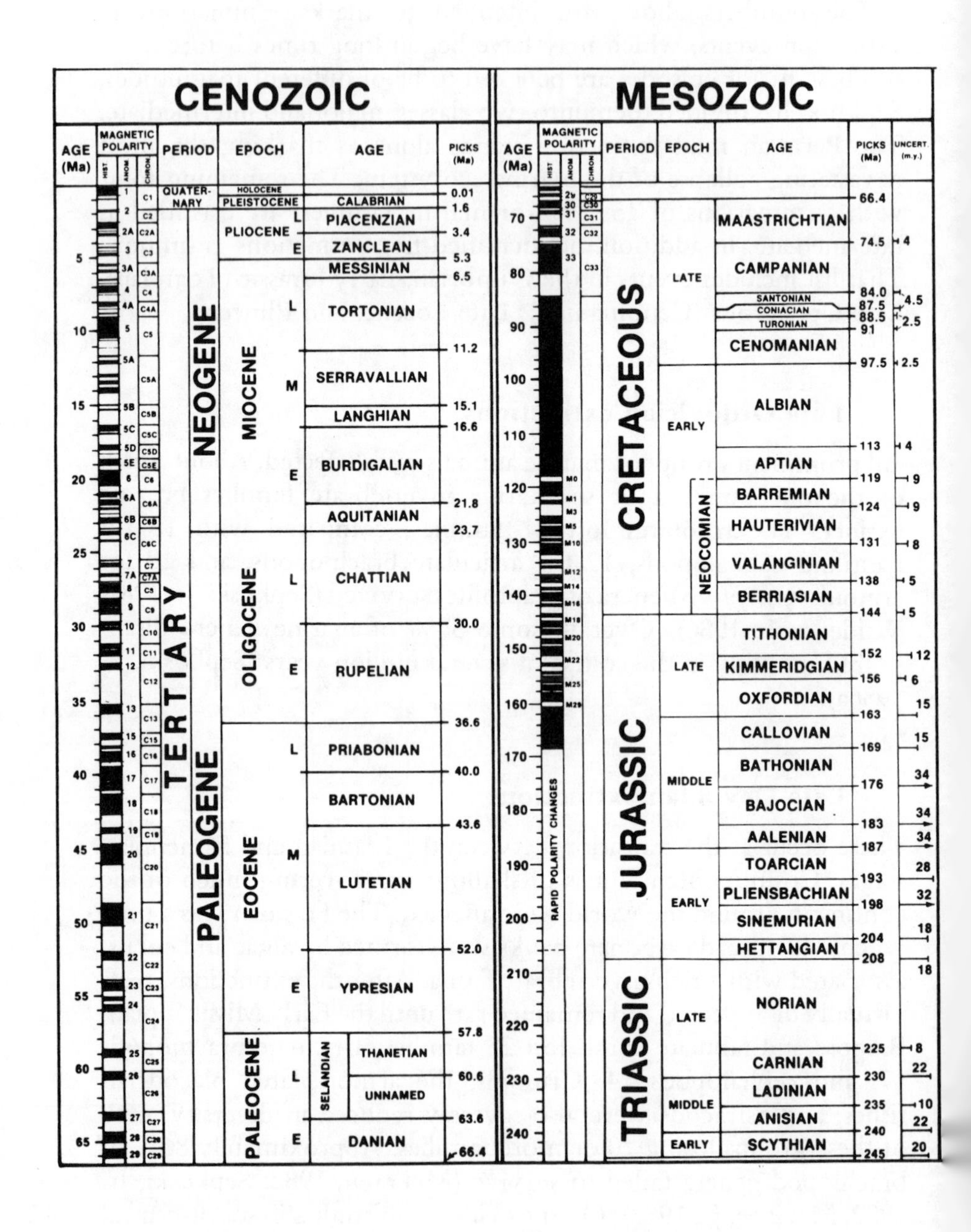
CENOZOIC
MESOZOIC
AGE (Ma)
MAGNETIC POLARITY
HIST
ANOM
CHRON
PERIOD
EPOCH
AGE
PICKS (Ma)
UNCERT. (m.y.)
QUATERNARY
TERTIARY
NEOGENE
PALEOGENE
HOLOCENE
PLEISTOCENE
PLIOCENE
MIOCENE
OLIGOCENE
EOCENE
PALEOCENE
CALABRIAN
PIACENZIAN
ZANCLEAN
MESSINIAN
TORTONIAN
SERRAVALLIAN
LANGHIAN
BURDIGALIAN
AQUITANIAN
CHATTIAN
RUPELIAN
PRIABONIAN
BARTONIAN
LUTETIAN
YPRESIAN
SELANDIAN
THANETIAN
UNNAMED
DANIAN
0.01
1.6
3.4
5.3
6.5
11.2
15.1
16.6
21.8
23.7
30.0
36.6
40.0
43.6
52.0
57.8
60.6
63.6
66.4
CRETACEOUS
JURASSIC
TRIASSIC
LATE
EARLY
MIDDLE
NEOCOMIAN
MAASTRICHTIAN
CAMPANIAN
SANTONIAN
CONIACIAN
TURONIAN
CENOMANIAN
ALBIAN
APTIAN
BARREMIAN
HAUTERIVIAN
VALANGINIAN
BERRIASIAN
TITHONIAN
KIMMERIDGIAN
OXFORDIAN
CALLOVIAN
BATHONIAN
BAJOCIAN
AALENIAN
TOARCIAN
PLIENSBACHIAN
SINEMURIAN
HETTANGIAN
NORIAN
CARNIAN
LADINIAN
ANISIAN
SCYTHIAN
66.4
74.5
84.0
87.5
88.5
91
97.5
113
119
124
131
138
144
152
156
163
169
176
183
187
193
198
204
208
225
230
235
240
245
RAPID POLARITY CHANGES

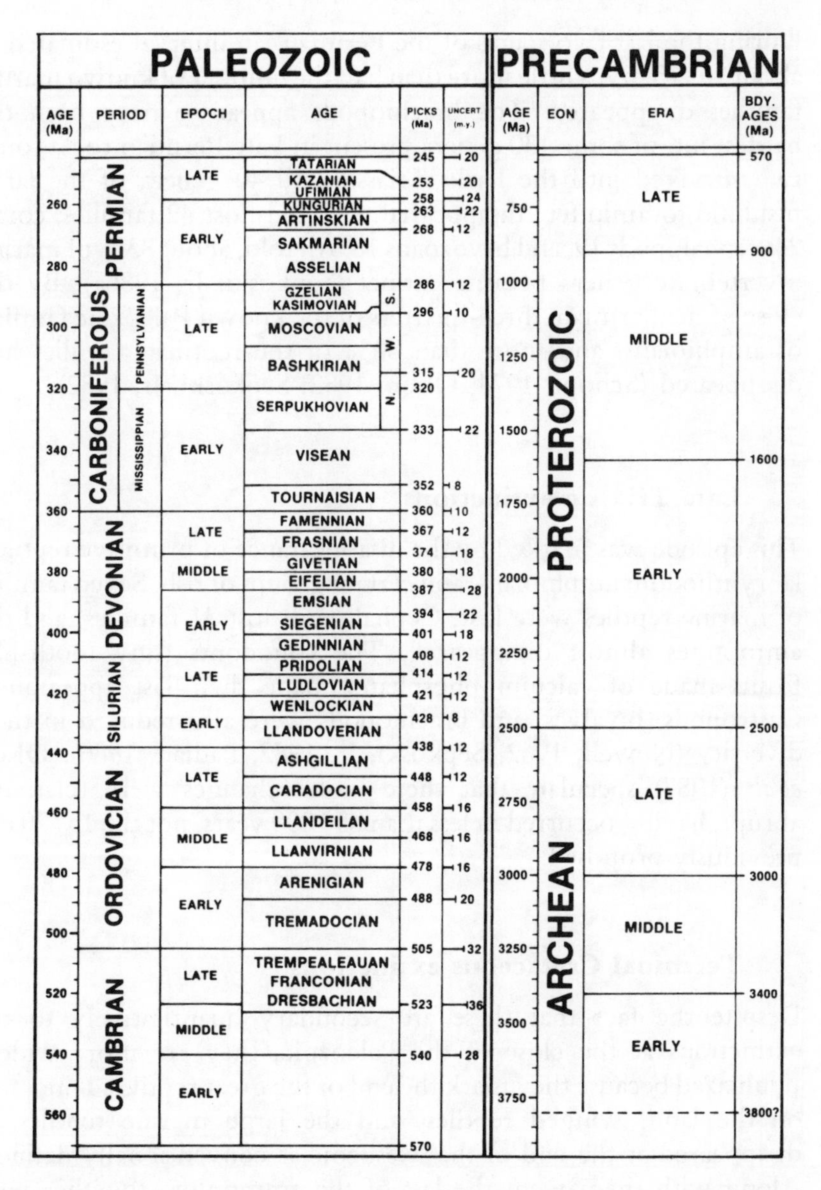

Figure 8.1 Geologic time scale. Compiled in 1983 by A. R. Palmer for publication by the Geological Society of America, and reproduced here with the permission of the Society.

Late Permian extinctions

During the last two stages of the Permian, an interval estimated at about 10–15 my, a little more than half the number of known marine families disappeared. The brachiopods appear to have been the hardest hit: of some 130 genera present in Late Permian rocks, only two survived into the Early Triassic. All 40 genera of the large fusulinid foraminifera disappeared. Crinoids lost 42 families, corals 24, cephalopods 19, and bryozoans 18. All told, about 83% of marine invertebrate genera became extinct (Sepkoski, Jr, 1986a). By the close of the Permian, three-quarters of the known Paleozoic families of amphibians, and more than 80% of the reptilian families had disappeared (Schopf, 1974; Olson, 1982; Sepkoski, Jr, 1982).

Late Triassic extinctions

This episode was marked by the disappearance of primitive reptiles, labrynthodont amphibians, and certain groups of fish. Seven families of marine reptiles were lost. Cephalopods lost 31 families, and the ammonites almost disappeared. The conodonts (tiny tooth-like fossils made of calcium phosphate) made their last appearance. Gastropods, bivalves, and brachiopods were also reduced in their diversity (Newell, 1962; Sepkoski, Jr, 1982; Padian, 1986). Olson *et al.* (1987) speculate that these biotic changes were relatively abrupt, having occurred in less than 850 000 years, not the 15–20 my previously proposed.

Terminal Cretaceous extinctions

Despite the fact that these are secondary quantitatively to the extinctions at the close of the Paleozoic, they are more widely publicized because they mark the end of the great reptiles. Dinosaurs on the land, winged reptiles, and the large marine reptiles all disappeared at the end of the Mesozoic as conventionally defined. Along with them went the last of the ammonites, together with whole families of gastropods, echinoids, large bivalves, fish, and much of planktonic marine life. All told, about 15% of known marine animal families became extinct (Sepkoski, Jr, 1982).

8.2 THE SEARCH FOR PERIODICITIES

Granting that these extinctions occurred at climactic times about where the approximate dates indicate, were they random or periodic events? In recent years the idea that mass extinctions have been periodic has gained increasing attention. The causes have been variously ascribed to terrestrial or extraterrestrial agencies, or to the conjunction of both.

A 50 million year cycle

That periodic bombardment of Earth by planetesimals reflects passage of the Solar System through successive spiral arms of the Galaxy at intervals of about 50 my is the premise for an hypothesis of mass extinction proposed by Napier and Clube (1979). Large objects, possibly as big as the moon, must abound in those spiral arms; and the asteroids within the Solar System may represent the residue of interstellar planetesimals accumulated during successive bombardment episodes.

During such a bombardment, 'blast waves' would preferentially exterminate the larger land animals. Fine dust exploded into the atmosphere would block sunlight and disrupt food-chains. With protracted cooling of the lands, continental glaciation might be initiated. The ozone layer would be depleted, and the deluge of ultraviolet radiation would be biologically devastating. Impacts of the cosmic buckshot in the oceans would raise waves up to 8 km amplitude, lands would be flooded, and obscuration of the upper atmosphere would speed the advent of ice-ages.

A 32 million year cycle

Fischer and Arthur (1977) have proposed that the extinctions are related to cyclic oscillations in the order of 32 my between two oceanic modes, which they designated as oligotaxic and polytaxic.

The present oceans exemplify the oligotaxic mode, which is associated with a major regression of the oceans. Past oligotaxic situations are characterized by a notable increase in extinctions of marine swimming and floating organisms. Biotic diversity diminished globally.

The polytaxic seas appear to have been associated with marine transgressions, and with times when the rates of speciation exceeded those of extinction. In comparison with the oligotaxic mode, the waters were warmer and had gentler thermal gradients both vertically and latitudinally. Oceanic currents convected sluggishly, in contrast to the more vigorous circulation of the oligotaxic mode. Bottom waters would become depleted in oxygen, favoring deposition of organic-rich black shale. That type of deposition would result directly and indirectly in a net loss of carbon from the hydrosphere and atmosphere. In the seas, the carbonate compensation depth (the CCD – that level below which dissolution of calcite exceeds supply) would rise.

Oligotaxic troughs coincide with at least the last seven crises in the history of life, as follows:

Boundaries	*Million years before present*	*Difference*
Permian–Triassic	220	
Triassic–Jurassic	190	30
Bathonian–Callovian	158	32
Early Neocomian	126	32
Cenomanian	94	32
Early Paleocene	62	32
Mid-Oligocene	30	32

Based on these figures, the Pliocene–Pleistocene extinctions, now underway, should reach a climax within the next two million years.

Aside from the paleontological evidence for the extinctions, the authors claimed that the oxygen isotope ratios indicate that relatively warm times coincided with polytaxic episodes, relatively cool times with oligotaxic biological crises. Systematic excursions in C^{12}/C^{13} ratios also fit very well into the 32 my cycle. Data from the Deep Sea Drilling Project in most instances reveal shoaling of the CCD during polytaxic times and deepening during the oligotaxic intervals. As an exception, a reduction of carbonate content occurs in basal Danian beds at the height of an oligotaxic episode.

Cores from oceanic sediments reveal extensive submarine hiatuses on most submarine rises and ridges and in some places on the abyssal floors, probably reflecting the action of strong currents in the deeps. These gaps correspond generally to oligotaxic episodes.

Fischer and Arthur did not profess to know the ultimate cause of these apparently periodic crises in the history of life. Since they link with oscillations in sea-level, however, they leaned toward a cause internal to the Earth rather than toward one related to solar or cosmic processes.

A 26 million year cycle

In 1984 Raup and Sepkoski offered statistical support for Fischer and Arthur's proposal that extinction events have occurred at regular intervals. Their original analysis indicated 12 extinction events from Permian to present, as opposed to the seven identified by Fischer and Arthur. Intervals between these were reduced from 32 to 26 my. Their more recent analyses have sustained the smaller interval, but the number of extinction maxima has been reduced to eight (Sepkoski, Jr, and Raup, 1986; Raup and Sepkoski, Jr, 1986; Sepkoski, Jr, 1986a). Based on figures for the time series of some 2160 families of marine animals, and nearly 11 800 of their genera (of which 9250 are extinct), these maxima appear as follows:

1. Late Permian (Guadalupian): most severe of all.
2. Late Triassic (Norian): one of five largest. Approximately 48% of marine genera disappeared (Sepkoski, Jr, 1986a).
3. Early Jurassic (Pliensbachian): precise timing and duration of event uncertain; perhaps a minor event.
4. Terminal Jurassic (Tithonian): 37% of marine genera disappeared (mostly ammonites, bivalves, and corals).
5. Early Late Cretaceous (Cenomanian): 28% of marine genera disappeared. On the evidence of stratigraphic studies in Colorado, Orth *et al.* (1987) proposed that extinctions of ammonites and benthic organisms proceeded in a series of five or six steps over an interval of about 3 my (see also Kauffman, 1986).
6. Terminal Cretaceous (Maastrichtian): second most severe; half of marine genera disappeared.
7. Late Eocene (Priabonian): 16% of marine genera disappeared during last 3.4–4 my of epoch. According to Kauffman, extinction proceeded in four steps.
8. Mid-Miocene (Langhian–Serravallian): most poorly documented of all, but in southern Germany there is an almost complete lack of larger reptiles, such as crocodiles and giant turtles, after the Ries impact–explosion event (Schleich, 1986).

To this list Sepkoski, Jr, (1986a) added a ninth mid-Early Cretaceous (Aptian) episode, during which 19% of marine genera disappeared.

The figure of 26 my for extinction periodicity holds very well for the last four and best dated events. But there are two gaps where extinction maxima should occur but apparently do not. One should fall in the Early Cretaceous, at some time between the Tithonian and Cenomanian; the other during the Middle Jurassic between the Tithonian and Pliensbachian. Sepkoski and Raup found these gaps troublesome, and wondered whether the forcing mechanism had 'missed a beat', or whether the gaps simply reflect our imperfect knowledge of extinction rates over the past 268 my.

As for the nature of the forcing mechanism, Raup and Sepkoski had no answer, though they believed it more likely to have been ultimately astronomical rather than terrestrial. They offered their findings simply as an hypothesis in need of further testing. Immediately this hypothesis stimulated renewed interest in possible astronomical mechanisms for mass extinctions.

A 30 million year cycle

As the Solar System orbits the Galaxy, it oscillates vertically through the galactic plane. The period of the oscillation is about 67 my. Thus approximately every 33 my the system passes through the galactic plane, and every 33 my it reaches one of the oscillatory extremes. Rampino and Stothers (1984) have proposed that this oscillation is the basic 'pace-maker' for mass extinctions.

After re-examining the data gathered by Raup and Sepkoski, they arrived at a figure of 30 ± 1 my as the interval between major extinctions of marine life during the last 250 my. As for the mechanism of extinction, they suggested that diffuse interstellar matter is concentrated in the galactic plane. As the Solar System passes up and down through that plane, the chances of collision with gas and dust of interstellar clouds would increase. If the Solar System should actually penetrate such a cloud, the Earth's upper atmosphere would be polluted with hydrogen gas.

Even if such a collision did not occur, the cloud's huge mass would gravitationally perturb the Solar System's inner reservoir of comets, and so 'focus a shower of comets on the innermost regions of the

Solar System'. This comet shower might last for 10 my, during which time one large and several small comets would hit the Earth, with disastrous effects on life.

Schwartz and James (1984) have also related mass extinctions to effects resulting from the sun's oscillation about the galactic plane. They accepted Rampino and Stothers' figure of approximately 30 my as the probable interval between successive extinctions of marine life. As for the mechanism, they proposed that if cosmic rays are confined by the magnetic field of the Galaxy, then the cosmic-ray flux would be less above plane than at mid-plane. Because cosmic rays influence the ionization balance in the upper atmosphere, there should be a decrease in the ionization rate as the sun approaches extremes in its oscillation, and a corresponding increase as it approaches mid-plane. Long-term changes in fluxes of cosmic rays could produce significant alterations in the biosphere, if not directly then indirectly by inducing cyclic changes in climate.

The Nemesis hypothesis

The possibility that the Raup–Sepkoski 26 my cycle is driven by the oscillations of a solar companion-star, as yet undiscovered, has been widely publicized in recent years. Travelling in a moderately eccentric orbit, this star at its closest approach would pass through the Oort Cloud of comets which surrounds the sun. During each passage the visitor would perturb the orbits of these comets, sending perhaps a million of them into paths which would invade the inner Solar System. Several of these should hit the Earth in the course of the following million years.

Presently the unseen companion is probably near its maximum distance from the sun, or about 2.4 light-years away. Therefore it should pose no further threat to life on Earth until around AD 15 000 000 (Davis, Hut, and Muller, 1984). In closing their argument, these authors proposed that 'if the companion is found, it should be named "Nemesis" after the Greek goddess who relentlessly persecuted the excessively rich, proud, and powerful. We worry that if this companion is not found, this paper will be *our* nemesis'.

A similar model, also based on the 26 my cycle, assumes a highly eccentric orbit for the companion star. Following each course through the comet cloud, there should be several terrestrial impacts.

These should occur over periods no longer than 100 000 to 1 000 000 years, as the fossil record requires that the duration of a shower should be less than a few million years (Whitmire and Jackson, 1984). For a chronological account of scientific speculations on periodicity of mass extinctions, see Raup (1986a).

Planet X

Calculations indicating that Pluto's mass is insufficient to account for discrepancies in motions of the outer planets have led to the speculation that an undiscovered tenth planet is responsible for periodic comet showers and mass extinctions (Whitmire and Matese, 1985). A belt or disk of comets beyond the orbit of Neptune was proposed as the source of the comets. Assuming that both this disk and Planet X actually exist, then during the life of the Solar System the planet must have swept out a clean path through the cometary disk. Widening of the path continues. During each precession of 58 my the planet approaches inner and outer margins of the path, triggering comet showers at intervals of one-half the precession period, or 28 my. These perturbed comets would be concentrated near the planetary plane. Their diffusion into the inner planetary system would lead to increased impacts on Earth, and thus possibly to periodic mass extinctions.

Periodicity in formation of cryptoexplosion structures

If the approximate ages in years of probable impact–explosion craters and structures were known, these might indicate whether or not their formation has been periodic. From a list of 88 features attributed to explosive impact, Walter Alvarez and R. A. Muller (1984) selected 13 whose age uncertainty was 20 my or less. They concluded that most of the ages for these fall in a 28.4 my cycle which, allowing for errors in measurement, is about the same as the Raup–Sepkoski interval between mass extinctions. With that figure Trefil and Raup (1987) are in substantial agreement. They have opted for a real cratering periodicity of about 29 my, while allowing that the majority of cratering events may be random rather than periodic.

Phanerozoic supercycles

Fischer (1984) has proposed that, beginning with the Cambrian Period, global climates have oscillated between the relatively warm and cold in response to long-term changes in sea-level and in the amounts of carbon dioxide and water vapor in the atmosphere. There have been two periods of exceptionally high sea-level: mid-Paleozoic and Cretaceous–Eocene. Widespread continental emergences occurred during the latest Precambrian, Permo–Triassic, and after the Late Eocene. These changes in the global eustatic curve he attributed to variations in the volume of the ocean basins.

During times when the continents were drifting apart, the spreading ridges would lengthen and their bulges would displace sea water, resulting in high sea-levels. Conversely, whenever the continents tended to reaggregate, lower levels would follow. Variations in the rate of spreading would also change the volume of the ocean basins. Fast spreading would produce broad and voluminous ridges, resulting in high stands of the sea. With decrease in spreading rate, the ridges would decay and the seas would retreat from the continents. In addition, eustatic shifts would attend changes in the thicknesses of the continental plates following episodes of rifting and reaggregation.

Volcanism is the ultimate source of carbon dioxide in the atmosphere and hydrosphere. Fischer assumed that the abundance of large granitic masses may be taken as an index of ancient volcanic activity. If so, volcanism evidently peaked during the early and mid-Paleozoic, and during the interval beginning with the end of the Triassic and ending in the Late Eocene. Release of carbon dioxide attending these eruptions would result in 'greenhouse climates', due to the thermal insulating properties of that gas. During the intervals between the greenhouse episodes, carbon dioxide would have been depleted and 'icehouse climates' would prevail. These colder climates, favorable for continental glaciation, appear to have coincided with the times of first-order low sea-levels noted above.

Fischer suggested that cross-overs between greenhouse and icehouse climatic regimes resulted in biotic crises. He cited the following four cross-over points.

Cross-over interval	*Approximate my ago*	*Intervals between cross-overs* (my)	*Change*
1. Late Cambrian	500		Icehouse to greenhouse
		145	
2. Late Devonian	355		Greenhouse to icehouse
		163	
3. End Triassic	192		Icehouse to greenhouse
		152	
4. Late Eocene	40		Greenhouse to icehouse

On the evidence of the above estimates, Fischer proposed that there have been two supercycles in prevailing global climates since Late Cambrian times. The first occurred over an interval of about 308 my, the time between cross-overs from icehouse to greenhouse conditions repeated in the Late Cambrian and terminal Triassic. The second began with transits from greenhouse to icehouse conditions initiated in the Late Devonian and repeated in the Late Eocene, an interval of about 315 my. As a mechanism, he suggested a cyclic process with a period of about 300 my related to plate activity and convection in the mantle.

Fischer conceded that his hypothesis does not account for all crises in the history of life, notably the major one at the end of the Paleozoic. Nor do reports of continental glaciation near the Ordovician–Silurian boundary fit into the greenhouse slot in his chronology. Perhaps the cause of this anomaly is to be found in some catastrophic extraterrestrial event, he speculated. As for the terminal Cretaceous extinctions, whatever the causes, they cannot be due to inversion of climatic state, for the Mesozoic greenhouse persisted into the Eocene.

8.3 SELECTIVITY IN EXTINCTION EVENTS

Whatever the causes of mass extinctions, it appears that some forms of life have been persistently more vulnerable than others. In his summary of a conference on causes and consequences of extinction Flessa (1986), identified the characteristics of the 'losers' as those endemic to geographically restricted areas, big-bodied taxa, taxa adapted to tropical habitats, taxa living in low-productivity habitats, taxa in evolutionary branches represented by few species, and taxa confined to terrestrial habitats.

From his analysis of casualties in the marine realm during the five major extinction episodes, McKinney (1987) concluded that planktonic and sessile benthic organisms have tended to suffer the highest mass extinction rates. The lowest extinction rates seemed to favor mobile benthic organisms such as bivalves, gastropods, foraminifera, and ostracods.

McKinney went on to propose that the essential difference between background and mass extinctions has been related to changes in the intensities and not the kinds of the processes responsible. He pointed out that the kinds of marine organisms that have suffered high rates of mass extinction have also exhibited high rates of background extinction. Conversely, those that have most persistently survived mass extinction episodes have also exhibited low background rates of extinction.

These views are contrary to some earlier interpretations of biotic changes in the succession of marine organisms. Raup and Sepkoski, Jr (1982) had concluded that at least for the Ordovician, Permian, Triassic, and Cretaceous events, the biotic patterns of mass extinction are statistically distinct from patterns of background extinction. Jablonski (1986a) had concurred. According to his analysis of the paleontological record, survivorship patterns have little correspondence with those of background times. In his words, 'lineages or adaptations can be lost during mass extinctions for reasons unrelated to their survival values for organisms or species during background times'.

9

Catastrophist scenarios for mass extinctions

9.1 THE ALVAREZ HYPOTHESIS FOR TERMINAL CRETACEOUS EXTINCTIONS

Near the Italian town of Gubbio, in the northern Appenines, the contact between Cretaceous and Tertiary strata of marine origin has been drawn precisely at a claystone parting only 1 cm thick. In the late 1970s a research team, organized at the University of California, Berkeley, set out to determine the length of time required for deposition of that boundary clay. Its members included Luis Alvarez, Nobel laureate physicist, his geologist son Walter and nuclear chemists Frank Asaro and Helen Michel.

Their approach was to determine the iridium content of the clay, and compare that with the estimated yearly infall of cosmic dust. Iridium was selected as the tracer element because it is about 10 000 times more abundant in stony meteorites than in crustal materials, and because modern analytic techniques can detect extremely small quantities of it.

Surprisingly, the iridium content of the boundary clay turned out to be too high to be explained in terms of average influx rates of cosmic matter. Even higher concentrations of the tracer were found along the Cretaceous–Tertiary (K–T) boundary at Stevns Klint, Denmark.

Late in 1979 the Berkeley team had formulated the hypothesis that the iridium spikes record the impact and explosion of an asteroid. This speculation gained international airing in 1980 after the publication in *Science* of their article boldly entitled *Extraterrestrial cause for the Cretaceous–Tertiary extinctions*. Few papers published in this century have generated so much controversy – or stimulated so many new investigations into the history of life on Earth.

The scenario of 1980 ran as follows. About 65 million years ago, an asteroid somewhere between 6 and 14 km in diameter struck the Earth and exploded. The energy thus released was probably the equivalent of 100 000 000 megatons of TNT. A vast cloud of dust, thrown into the air, shut out the sunlight, and turned day into night for a period of several years. During this time of darkness, photosynthesis was suppressed, and food chains collapsed. In the marine realm microscopic floating plants and animals were nearly exterminated. On land, plants died, large herbivores starved, thus depriving the large meat-eaters of their rations. When the light returned, plants that could reproduce from seeds, spores and roots revived. Animals that could feed on insects and decaying vegetation likely survived the ordeal.

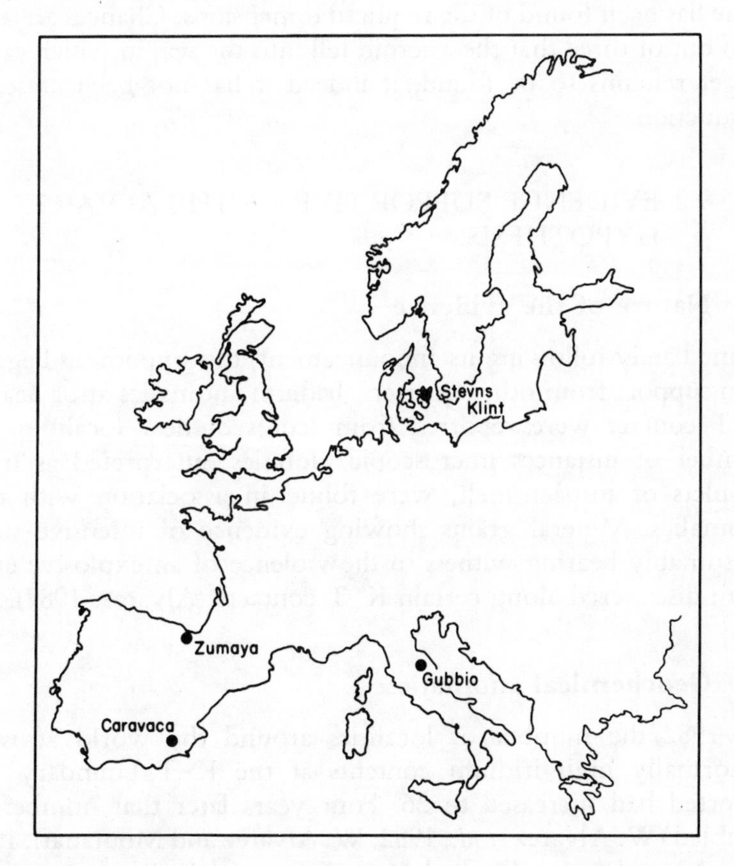

Figure 9.1 Map showing localities in Europe where the K–T boundary has been examined intensively.

As prime evidence for the impact, the authors cited the iridium spikes in boundary sediments. At Gubbio the background count of 0.3 parts per billion (ppb) for the uppermost Cretaceous increases abruptly to 9 ppb at the boundary in one section. At Stevns Klint the iridium in the boundary clay rises over background level by a factor of 160. No reasonable alternatives could be found for the proposition that there was an abnormal influx of extraterrestrial material during the K–T transition. The possibility that this influx had been related to explosion of a supernova was discounted, because such an event should have brought plutonium to the boundary layer along with the iridium. In fact, no plutonium showed up in the analyses (Figure 9.1).

The impact should have produced a sizeable crater, but thus far none has been found of the required dimensions. Chances are about two out of three that the asteroid fell into the sea, in which case its target remains to be found, if indeed it has not been erased by subduction.

9.2 EVIDENCE SUPPORTIVE OF THE ALVAREZ HYPOTHESIS

Nature of the evidence

Immediately following its announcement, this hypothesis began to gain support from other quarters. Iridium anomalies at or near the K–T contact were reported from scores of new localities. In a number of instances microscopic globules, interpreted as frozen droplets of impact melt, were found in association with those anomalies. Mineral grains showing evidence of intensive shock, presumably bearing witness to the violence of an explosive event, were discovered along certain K–T contacts (Alvarez, 1987).

Geochemical anomalies

By 1982 the number of localities around the world at which abnormally high iridium contents at the K–T boundary were reported had increased to 36. Four years later that number had doubled (W. Alvarez *et al.*, 1982; W. Alvarez and Montanari, 1985). The original anomalies had been discovered in European marine sequences. Thereafter they have been found in marine deposits of other continents, in cores of deep-sea sediments and at exposures of continental deposits.

As analyses of boundary sediments progressed, exotic noble elements other than iridium were discovered. At a locality in Denmark, marine sediment at the K–T transition was found to contain traces of osmium, gold, platinum, rhenium, ruthenium, and palladium, all five or more times higher than their expected abundances (Ganapathy, 1980). Near Caravaca, Spain, the K–T contact in marine sediments has been drawn along a layer of clayey marl 10 cm thick. In the basal few centimeters of that parting, iridium and osmium are present in amounts respectively 450 and 250 times normal (Smit and Hertogen, 1980).

Soon after the iridium anomalies were discovered in Europe, the search for these began in other continents. Analyses of boundary sediments in a sequence of marine deposits exposed along the Brazos River in Texas showed that the content of iridium increased about fifty-fold over background to a level of 2.10 ppb at the contact (Ganapathy *et al.*, 1981; Hansen, 1984). Coarse-grained sediments along that contact have been interpreted as the work of a tsunami, probably generated by impact of a bolide (Bourgeois *et al.*, 1988). The first discovery of an anomaly in the Southern Hemisphere at a land-based site was recently confirmed for certain shale deposits in New Zealand (Brooks *et al.*, 1986).

As long as the known iridium anomalies were confined to marine sediments, the possibility existed that these were due to processes of concentration peculiar to the oceanic realm. That speculation was discredited with the discovery of anomalies in sediments of continental origin in the Western Interior of the United States. In the Raton Basin of northern New Mexico and southern Colorado, the K–T boundary has been defined on the basis of an abrupt drop in the ratio of angiosperm pollen to fern spores. At two localities 50 miles (80.47 km) apart, an iridium anomaly is associated with this botanical discontinuity. Concentrations of iridium range upward to 5 000 ppt (parts per trillion) against a background of 4–20 ppt (Orth *et al.*, 1981, 1982). A similar situation has been reported farther north in Wyoming and Montana, where the palynological K–T boundary clay layers contain anomalously high concentrations of the trace element (Tschudy and Tschudy, 1986). At a locality in Wyoming, the contact between the Late Cretaceous Lance Formation and the Paleocene Fort Union Formation is marked by a boundary clay a few millimeters thick containing an iridium spike of 21 ppb (Bohor, 1987). The clay also contains shock-metamorphosed mineral grains and microspherules. A fragmental dinosaur bone was recovered about 1 m below the contact.

Microspherules

Microscopic particles, variously globular or discoid in shape, have been found in boundary sediments at an increasingly large number of localities. At Caravaca these range between 0.2 and 1.3 mm in diameter and consist almost entirely of potassic feldspar containing traces of iridium measured at 10.4 ± 0.6 ppb. Here they are unusually abundant, numbering 50 to 300 particles per cm^3 (Smit and Klaver, 1981; Montanari *et al.*, 1983). Spheroids of feldspar, glauconite, and magnetite have been reported from 16 sections in the northern Appenines, in numbers ranging upward to 230 per cm^3 (Montanari *et al.*, 1983).

Shocked mineral grains

Mineral grains showing evidence of shock-metamorphism appear to be common constituents of the K–T boundary beds in the Western Interior of North America. At 20 sites, scattered from Alberta, Canada, to New Mexico, these grains are concentrated in laminated clay 0.5–0.8 cm thick. Clay-free residue consists mainly of quartz, feldspar, and grains derived from siliceous rocks. Except for the grains of chert, about 30% of them show the intersecting planar features characteristic of shocked materials. At nearly all localities, coal beds of Tertiary age overlie the units containing these grains. These occurrences are notable in that the shocked particles range upward to more than 0.5 mm across, the largest yet reported elsewhere in the world. Because of their abundance, coarseness, and composition, Izett and Bohor (1986) concluded that the postulated impacting body struck quartz-rich rocks somewhere in North America.

More recently, shocked quartz grains have been reported from boundary clays containing iridium 'spikes' at five sites in Europe, in a core from the north-central Pacific Ocean, and at a locality in New Zealand. Whereas the first discovery of shocked quartz along the K–T boundary in Montana was within a sequence of continental deposits, all seven of these finds were in beds of marine origin.

Bohor, Modreski, and Foord (1987) have taken this as confirmation of the Alvarez hypothesis, holding that 'an impact event at the Cretaceous–Tertiary boundary distributed ejecta products in an earth-girdling dust cloud'. Moreover, they expect that shocked minerals are to be found in every K–T boundary clay that contains a substantial iridium anomaly 'anywhere in the world'.

9.3 CONSEQUENCES OF IMPACT AND EXPLOSION OF EXTRATERRESTRIAL BODIES

Cometary impact in the ocean

In the same year that the Alvarez hypothesis gained international attention, Kenneth Hsü proposed cometary impact as an alternative to impact by an asteroid. He assumed that a comet large enough to form a crater several hundred kilometers in diameter had dived into the ocean. Resulting short-lived changes in the temperature and composition of the hydrosphere and atmosphere could account for the terminal Cretaceous extinctions, he proposed. His original scenario ran as follows.

Heat, generated during the comet's transit through the atmosphere and by its subsequent explosion, may have risen to levels sufficient to induce nuclear or even thermonuclear reactions. Air temperature possibly increased by 10° to 20°C. Large land vertebrates perished under almost instantaneous thermal stress. Small or aquatic terrestrial animals largely survived. Turtles could escape the heat wave by diving under water and holding their breath. Crocodiles survived by burying their eggs in mud. Land vegetation could have escaped extinction as new seeds sprouted.

Meanwhile, dissolution of the cometary nucleus in the ocean should have released a large quantity of poisonous substances, such as hydrogen cyanide (HCN), methyl cyanide (CH_3CN), and possibly carbon monoxide (CO). Dissolved in sufficient concentrations, these would kill all kinds of marine organisms. In due course, the cyanides would be reduced through oxidation to produce carbon dioxide (CO_2). Added to the CO_2 brought in by the comet, the acidity of the ocean would increase. Calcareous planktonic life would be virtually extinguished. Mass mortality of the phytoplankton, the basis for food chains serving higher organisms, would lead to starvation of marine animals that had escaped cyanide poisoning.

Molten rock in the suboceanic impact crater should have

generated additional thermal stress, and so promoted active circulation of bottom waters – hence the depositional hiatus at many deep-sea drilling sites. Earthquakes and tsunamis should have caused subaqueous slumping and deposition of turbidites.

In his more recent writings, Hsü granted that the question as to the nature of the impacting body, whether cometary or asteroidal, remains open. In either case the warming of the ocean was not directly due to impact. He withdrew his proposal that oceanic plankton died from cyanide poisoning: all the cometary cyanide should have been decomposed by heat and shock pressure. Flash-heating of the atmosphere he abandoned as the cause of extinctions, except in 'disaster areas' around the site of impact. However, he has held to the idea that for a time after impact the oceans were more acid, or at least less alkaline than normal. In addition to the acidity contributed by increased content of carbon dioxide, combination of nitrogen and oxygen during passage of the bolide through the atmosphere should have produced nitric oxide, which would transform to nitric acid and fall as acid rain. The large increase in atmospheric carbon dioxide probably produced a greenhouse effect that may have lasted as long as 50 000 years, as indicated by carbon and oxygen isotope ratios (Hsü, McKenzie, and He, 1982; Hsü *et al.*, 1982; Hsü, 1984, 1986).

Cometary or asteroidal impact in the ocean

A third hypothesis for the K–T extinctions envisioned an asteroid or comet nucleus plunging into the northern Pacific Ocean (Emiliani, Kraus, and Shoemaker, 1981; see also Melosh, 1982). Assuming that the object involved was about the same size as called for according to the Alvarez hypothesis, the sequence of events was pictured as follows.

In approaching its target, the bolide opened a hole in the atmosphere which expanded to a radius of several hundred kilometers before the air could flow back and fill it. A powerful wind, capable of flattening forests to distances between 500 to 1000 km, blew outward for about an hour. Then there was a strong return wind.

On impact with the ocean, a water layer was dispersed over a circular area possibly with a radius of at least 100 km. Water in the target area was first compressed, then vaporized.

Assuming an ocean depth of 5 km and that the ocean crust beneath it was also 5 km thick, then about 35% of the transient cavity was excavated in water, 25% in oceanic crust, and 40% in the underlying mantle. Monstrous gravity waves with heights of several hundred meters travelled thousands of kilometers across the ocean. Super-tsunamis penetrated far into the surrounding continents, possibly exterminating species whose habitats were restricted to coastal lowlands.

About half the volume of shocked rock remained in or settled back into the transient seafloor cavity. Ocean water flowed back. More slowly, material from the upper mantle flowed toward the transient cavity, flattening the floor of the initial crater. As the floor rose, the rim subsided. Scarps formed around the crater in the seafloor.

Meanwhile a plume of steam and volatilized, melted and solid rock had been sprayed into the atmosphere. Much of this hot stuff was trapped in the stratosphere. The global surface temperature increased by 0.5° to 2°C. This was followed by a cooling episode over extensive areas as the spreading dust cloud intercepted the sun's rays. Most of the dust, however, was washed out within weeks or months. The remaining water vapor and clouds produced a greenhouse effect, resulting in a substantial rise in global surface temperatures, perhaps by 10°C or more over a period of months or years. This thermal stress was particularly hard on reptiles, if indeed those of the Late Cretaceous, like living reptiles, could withstand considerable drops in temperature but could not survive rises of a few degrees above temperatures optimal for their vital activities.

Possibility of mass extinctions caused by comet showers

That there may be a causal connection between comet showers, clusters of impact–explosion events and stepwise mass extinctions that may extend over intervals of one to three million years is an hypothesis cautiously proposed by Hut *et al.* (1987). Comet showers might be triggered during passages of neighboring stars near the cloud of comets (Oort Comet Cloud) surrounding the sun. In the course of such perturbations, comets would shower into the inner Solar System. Thus multiple impacts over protracted periods of time might account for the stepwise mass extinctions proposed for the Cenomanian–Turonian, Cretaceous–Tertiary and Eocene–Oligocene transitions (Donovan, 1987).

Wildfires

Global wildfires have recently been proposed as a mechanism for extinction, as well as for abrupt changes in flora at the K–T boundary (Wolbach, Lewis, and Anders, 1985; Saito, Yamanoi, and Kaiho, 1986). Samples from boundary clays in Denmark, at Caravaca, and at Woodside Creek, New Zealand, have been found to contain 0.36–0.58% of microscopic particles consisting mainly of elemental carbon. These have been interpreted as 'soot' from a global conflagration ignited by impact and explosion of a giant meteorite.

The proposed connection between soot and extinction runs as follows. Considering the close proximity of the continents to each other during the Late Cretaceous, a bolide exploding, say, in the Bering Sea could ignite vegetation over a radius of more than 1000 km. Thus both Eurasia and North America would have been within ignition range. Once started, the fire could spread over entire continents, burning standing vegetation as well as organic matter in the topsoil. Terrestrial animals would be devastated, if not by flames then by toxic gases such as carbon monoxide. The black smoke ascending into the atmosphere would absorb sunlight more effectively than rock dust, and so would quickly block photosynthesis. Until nearly all the soot had settled, the Earth would experience a cooling like that envisioned for a 'nuclear winter'. Perhaps much of the soot came from ignition of fossil carbon rather than from living organisms. In any case the net effects on the biosphere, while disastrous, would not have been apocalyptic since about half the known Cretaceous genera managed to survive.

Publication of this hypothesis stimulated proposals for different causes of the fires and other sources for their fuels. Wildfire moves slowly through green forests, and is usually terminated in hours or days by changes in wind direction or by rain. Suppose, however, that prior to the burning, impact and explosion of an asteroid had raised a dense and world-wide dust cloud. Night-time temperatures would have been depressed, possibly by 10° to 20°C. The result would have been the killing of tropical rain forests, perhaps of other forests as well. Lightning strokes – principal causes of wildfire until *Homo pyrotechnicus* appeared on the scene – would have set this dead vegetation aflame on 'every treed continent in both hemispheres, ensuring world-wide distribution of smoke sources' (Argyle, 1986).

The possibility that much of the soot came from ignition of fossil carbon was enlarged upon by Cisowski and Fuller (1986). They pointed out that spontaneous combustion in marine sediments recently exposed to subaerial weathering is a fairly common phenomenon. For example, at the time of their writing, marine black shales of Middle to Late Cretaceous age were burning at numerous places in north-western Canada near the Beaufort Sea, as well as in western Greenland. Other historically active combustion–metamorphism sites are known to have burned for decades. Some croppings with histories of intermittent burning have reignited over centuries. With widespread retreat of the seas during Late Cretaceous time, combustion-prone bituminous materials would have been widely exposed to subaerial erosion. Perhaps their ignition was triggered by a meteoritic explosion.

Immediate and long-term effects of impacts

Since 1981 there have been numerous speculations concerning immediate and long-term effects attending impacts of asteroids and comets striking land or diving into the ocean.

In the case of an impact on land by an asteroid, O'Keefe and Ahrens (1982) calculated that a mass 10 to 100 times the mass of the bolide would be injected into the atmosphere. Bolide-enriched ejecta would consist of condensed vapor, melt droplets and finely comminuted solid particles. Lofted to heights of at least 10 km, millimeter- to centimeter-size droplets would re-enter the atmosphere to form tektites and microtektites. Smaller particles would rise to even greater heights and be distributed globally in a matter of months. Photosynthesis would be reduced.

Due to heating by ejecta, a significant portion of the atmosphere and the upper one meter of the ocean could rise by about 15°C in temperature immediately following impact. Then, due to the shielding effect of atmospheric dust, there would be a world-wide cooling period lasting 10 to 100 days. This would be most severe on the continents. At sea, lack of sunlight would halt photosynthesis in phytoplankton and lead to the collapse of the marine food chain.

Finally, the water and/or nitric oxide in the ejecta might give rise to a terrestrial greenhouse. In addition these substances could deplete ozone in the stratosphere and mesosphere, and so release life-damaging ultraviolet irradiation.

If the asteroid struck the ocean, an additional mass of 10 to 100 times that of the bolide would be ejected into the atmosphere. As in the above case, there would be a surge of heating in a significant part of the atmosphere and upper one meter of the oceans, this time by about 5°C. That would be followed by days of cooling due to dust and cloud decks, succeeded by a more protracted greenhouse effect.

In the case of a cometary impact, whether on land or water, a large part of the ejecta would consist of water, according to O'Keefe and Ahrens. Much of the water would condense and fall as rain. Enhanced greenhouse effects over the continents and decreased concentrations of ozone above would be the probable consequences.

Whether shock-heating of the atmosphere were due to the impact of a comet or an asteroid, the result would be addition of large quantities of nitrogen oxides, initially nitric oxide (NO) (Lewis *et al.*, 1982). This compound would quickly convert to NO_2, which ultimately would form nitrous (HNO_2) and nitric (HNO_3) acids. The rainout of these acids, along with others, would acidify surface waters, selectively destroying shells made of calcium carbonate.

During the weeks or months of severe acidic pollution, seedlings would be injured and more mature continental plants defoliated. Terrestrial animals most likely to survive would probably include burrowing creatures with food hoards in alkaline soils. Action of the acids on carbonate substances would release carbon dioxide into the atmosphere, thus enhancing the greenhouse effect.

According to Prinn and Fegley, Jr (1987), the impact of an ice-rich long-period comet with a mass of 1.25×10^{16} kg and a terminal velocity of 65 km/s would produce enough nitric acid, acid rain, and NO_2 to traumatize the biosphere on a global scale. Calcareous organisms in the top 150 m of the world ocean would be annihilated. Massive chemical weathering of continental soils would release toxic metals to streams, whose waters would further pollute the oceans. With defoliation of the continental flora, the large herbivores would starve. Organisms most likely to survive would include those inhabiting the deep oceans, upper-ocean organisms with siliceous skeletons, the fauna inhabiting lakes buffered by carbonate rocks, and animals living or laying eggs in burrows excavated in well-buffered soils.

Macdougall (1988) has also suggested that enhanced chemical weathering of continental rocks induced by acid rain attending a large bolide impact may have been responsible for anomalously high concentrations of strontium reported in marine strata near the K–T boundary. He pointed out, however, that increased volcanic activity around the end of the Cretaceous could also have led to acid precipitation, due to increased emissions of gases transforming to sulfuric and hydrochloric acids.

The fact that no crater or structure of the size predicted by the Alvarez hypothesis has yet been found, has led to the proposal that bolides of lesser diameter than 10 km could account for the extinctions (Weissman, 1982; Padian, 1984). Model calculations suggest that hypervelocity impacts by an asteroid only 0.4–3 km in diameter, or a close encounter with a Halley-sized comet, would produce an aerosol mass sufficient to account for the darkness scenario (Gerstl and Zardecki, 1982; Padian *et al.* 1984).

Possibility of multiple bolides related to Cretaceous and Devonian extinctions

Alternatively, the impact may have been delivered, not by a single large object, but by a swarm of smaller bodies. McKay and Thomas (1982) have suggested that an encounter with a blast of cosmic buckshot, lasting for about a year, might produce the effects envisioned in the Alvarez hypothesis. The authors emphasized, however, that 'there is no astrogeophysical evidence that such swarms exist in the solar system'.

McGhee, Jr (1982) has called attention to the fact that five probable impact structures have been assigned dates that fall near the 362 ± 6 my estimated for the Late Devonian Frasnian–Famennian boundary.

Name and location of structure	*Estimated age* (my)	*Diameter* (km)
Charlevoix, Canada	360 ± 25	46.0
Keluga, USSR	360 ± 10	15.0
Flynn Creek, USA	360 ± 20	3.8
Siljan, Sweden	365 ± 7	52.0
Crooked Creek, USA	320 ± 80	5.6

He suggested that dust clouds raised by a succession of impacts such as these may have led to global climatic cooling responsible in turn for mass extinctions of Late Devonian marine organisms intolerant of cold water.

Evidence for climatic change provided by tektites

Glass (1982) has called attention to the possible correlation between impact events and climatic cooling based on evidence provided by tektites. Of the four areas on the Earth's surface where tektites are widely dispersed, two have yielded specimens with dates roughly coinciding with times some have claimed as marking crises in the history of life.

Specimens from the Czechoslovakian strewnfield have been dated at 14.7 my (Middle Miocene). Those from the North American strewnfield date from the Late Eocene, about 34 my ago. Glass speculated that the Miocene event may have precipitated the drop in temperature that led to the Earth's present glacial mode. 'Within the limits of resolution', he concluded, 'it seems possible that the impact events that produced the tektite strewnfields triggered climatic changes which in turn affected the marine biota in various degrees.'

9.4 RADIATION AS A CAUSE OF MASS EXTINCTIONS

Supernovae

Schindewolf (1963) was an early proponent of the idea that large and acute doses of cosmic radiation might have accounted for mass extinctions. Supernovae have been identified as the most likely source of lethal radiation.

Nova is the name astronomers give to a star that increases in luminosity 100 to 1 000 000-fold over a period of a few days. A supernova may be about 100 times brighter; and at its brightest it may be more than a thousand million times more luminous than our sun. High luminosity may persist for a week or two, after which a gradual decline in brightness follows over a period of years. In the course of its explosive phase, the star appears to have lost a large part of its mass, and in that process to have generated a burst of cosmic rays. In historic times at least three supernovae have appeared in the Earth's Galaxy: Tycho's Nova (1572), Kepler's Nova (1604), and the Crab Nebula (1054, according to Chinese records). Astronomers reckon that the explosion that produced the Crab Nebula occurred about 5000 light years from the Earth. The nebula continues to emit radiation ranging in wavelength from radio waves to X-rays.

Experiments conducted at Brookhaven National Laboratory, New York, showed that North American forests are damaged by exposures to radiation at the same levels as those approaching lethal levels in humans exposed to radiation from nuclear explosions. Pine trees are especially sensitive; they begin to suffer ill effects after six month's exposure to radiation of only 1 or 3 roentgens (r) per day (about the same level as the X-rays administered at the dentist's office). When the level was raised to 20 or 30 r per day, the trees died after six months' exposure. No oaks or pines survived when daily levels were raised above 350 r. Weeds were more hardy, surviving protracted daily exposures up to several thousand roentgens, while mosses and lichens survived total exposures of more than 200 000 r (Woodwell, 1963, 1967).

Net accumulation of litter and humus provides a useful index for measuring the productivity of a forest. The Brookhaven experiments showed that production of litter in an oak–pine forest showed a progressive decline after exposure for twelve years to gamma radiation. After radiation ceased, new and more resistant species of plants invaded the forest (Armentano and Woodwell, 1976).

Direct effects of supernova radiation

If a supernova explosion were to occur less than 100 light years from the Earth, it has been estimated that the burst of radiation would impact the planet over a period of a few days at most. Making what they considered reasonable assumptions as to the time intervals between 'nearby' explosions, and as to variations in their magnitude, Terry and Tucker (1968) estimated that the Earth has probably been exposed to an acute dose of 500 r every 50 my, and of 1500 r every 300 my. Experiments with laboratory animals were cited to indicate that about half those tested died after exposures to 200–700 r. Most of the survivors were rendered sterile. They concluded that heavy doses of radiation could account for mass extinctions of many exposed animals, including some shallow-water aquatic organisms, without simultaneous extinction of plant life.

Indirect effects of supernova radiation

Other authors would attribute the damage to life inflicted by supernovae not directly to the initial radiation flashes, but to the effects they would have on the ozone (O_3) that absorbs solar ultraviolet radiation in the stratosphere. The flash would lead to dissociation of nitrogen. Free nitrogen atoms would quickly oxidize to form nitric oxide, which in turn would combine with the ozone and convert it to O_2. Without its protective ozone 'umbrella', life on Earth would be subjected to lethal solar radiation.

Ruderman (1974) has suggested that more than 90% of the ozone may have been destroyed by this mechanism at least a few times since life first appeared. He estimated that this depletion may have lasted for intervals of a few years to a century.

The term 'nearby supernova', as employed by astronomers, may be misleading to others. By 'nearby', Clark, McCrea and Stephenson (1977) meant something closer than 3262 light years, or about 30 000 000 000 000 km. These authors proposed that the solar system should encounter one of these violent neighbors on an average of once every time it goes around the Galaxy; and since ozone contributes to the greenhouse effect, its depletion could lower global temperatures and thus possibly initiate an ice-age.

The possibility that the Earth may at some time have actually passed through a supernova remnant-shell has been raised by Reid *et al.* (1978). They speculated that this passage 'would cause a long period of harsh environmental conditions for the biosphere'. Increased levels of ultraviolet radiation, a cooler and drier climate, and reduced photosynthetic activity are among the factors that might cause mass extinctions.

Sporadic solar flares

Solar flares appear as unusually bright spots in the sun's lowest atmosphere. They may develop in a few minutes and last for several hours. Flares emit ultraviolet radiation, X-rays and energetic particles such as protons. Particles reach the vicinity of the Earth in a day or two. On their way they may subject human beings in space to hazardous radiation. On impacting the Earth's magnetic field and ionosphere, they cause auroras and disrupt radio communication.

Abnormally energetic flares might be harmful to life in two ways. Ionization in the stratosphere would lead to depletion of ozone. If the flare appeared during a time when the Earth's magnetic field was in the process of weakening or disappearing – say over a period of a few thousand years – the influx of solar and cosmic energy might lead to mass extinctions (Reid *et al.*, 1978; see also Uffen, 1963 and Cox, 1969).

Increased radiation attending magnetic reversals

Citing evidence from growth rings on Devonian corals to indicate that there were, some 390 my ago, 400 solar days in the year, Whyte (1977) proposed that secular decelerations in the spin of the Earth on its axis have been interrupted by episodes of acceleration. During the turning points from maxima and minima in rate of spinning, there would be periods of instability in the Earth's magnetic field, when the biosphere would be subjected to increased doses of radiation. Turning points from minimal to faster rates were equated with mass extinctions in the Late Ordovician and Permian. Turning points from maximum to slower rates were correlated with Late Devonian and terminal Cretaceous extinctions.

Individual magnetic reversals, as presently identified, do not show a close correspondence with times of mass extinctions. However, McCrea (1981) has noted that two major extinctions, the terminal Permian and Cretaceous, occurred during intervals when there were many magnetic reversals. If increased influxes of cosmic rays during reversals are indeed capable of producing significant biological effects, then ray-bombardment during any one in a cluster of these reversal episodes was potentially capable of adversely affecting organisms for which some crucial instinctive behavior was linked to the geomagnetic field.

That impacts and explosions of extraterrestrial bodies may have been responsible for magnetic reversals as well as for extinctions of species is a possibility recently explored by Muller and Morris (1986). (Also see Schwarzschild, 1987.) The sequence of events in their scenario runs as follows.

A large asteroid or cometary nucleus strikes the Earth and explodes. Dust lofted into the atmosphere brings on a 'nuclear winter'. The cold persists long after the dust has settled, because of the increased reflectivity of the snow-covered continents. In the course of a few centuries, equatorial water is transported to form polar ice caps, so that global sea-level drops by about 10 m. As a result of that transfer in mass, the moment of inertia of the solid and outer reaches of the planet is reduced by about one part in a million. Mantle and crust are set spinning faster than the solid iron inner core at the center of the Earth. As a result, the 2300 km-thick shell of liquid outer core separating the mantle from the inner core acquires a velocity shear. In the course of about a thousand years, this shearing destroys the pattern of convection that served as a dynamo maintaining the Earth's dipole field. The dipole field is destroyed. Perhaps 10 000 years later a new stable dynamo field would establish itself.

Muller and Morris acknowledge that no magnetic reversal is known at the K–T boundary, but that does not invalidate their hypothesis. The terrestrial dynamo has no memory. Thus the chances are 50–50 that a new magnetic field would have the same polarity as its predecessor.

Courtillot and Besse (1987) would relate variations in frequency of reversals of the geomagnetic field to variations in the rate of polar wander. Polar wander involves the shifting of the entire mantle relative to the Earth's axis of spin. These changes would affect thermal and chemical coupling in the Earth's multi-layered heat machine – core, mantle and lithosphere. During times of polar standstill, there would be a decrease in reversals. Two exceptionally long intervals when the magnetic field failed to reverse came near the ends of the Paleozoic and Mesozoic eras – a span of about 70 my for the Permo–Carboniferous, and about 35 my for the Cretaceous. In both instances these episodes of geomagnetic stability ended in episodes of violent volcanism. The authors theorize that times of minimal instabilities in the axis of the Earth's spin and in the geomagnetic field would be attended by maximum heat flow into the mantle from the core. Ultimately, a surge of thermal energy would ascend toward the Earth's surface, and there generate exceptional volcanic activity that might disrupt global climates and thus lead to mass extinctions of organisms.

9.5 'BAD WATER' HYPOTHESES

Arctic spillover

Mass extinctions at the close of the Cretaceous have been attributed to the sudden emptying of cold, fresh or brackish water from a previously isolated Arctic Ocean. In the Late Maastrichtian, according to this hypothesis, the Arctic became isolated from the world ocean as a result of crustal movements attending withdrawal of the epicontinental seas and partial filling of them with sediment. Salt water was flushed out by excess precipitation and runoff. Thus was the Arctic Ocean transformed into a large body of fresh or brackish water.

Probably as a result of rifting between Greenland and Norway, a connection was again established with the North Atlantic. Denser Atlantic water of normal salinity intruded the Arctic Basin, filling it from the bottom upward. The lighter fresh or brackish water spilled out and covered the entire world ocean with a lethal blanket of low-salinity water. Spillover and spreading probably lasted several years. The layer of sub-saline water was probably thick enough to occupy the entire photic zone.

Planktonic organisms intolerant of reduced salinity perished. Many species of shallow-sea benthonic organisms suffered the same fate. In the surface waters, now depleted in oxygen as well as salt, food chains were disrupted, and the ammonites and large marine reptiles disappeared.

Spillover of cold Arctic water caused a drop of about 10°C in the surface temperature of the oceans. The atmosphere chilled. Reduced evaporation over the oceans resulted in reduced precipitation over the lands by about a half. Onset of intense drought, perhaps lasting no more than a decade, combined with increased fluctuations in seasonal temperatures, led to a reduction of tropical and subtropical flora and to a great expansion of savanna-type vegetation. The climatic change was so abrupt as to impact communities of cold-blooded reptiles; and the radical change in vegetation disrupted food supplies for all herbivores. Small and versatile mammals, native to temperate climate and savanna vegetation, moved in to compete for room and food in territories formerly dominated by dinosaurs. Reptiles that survived, such as crocodiles and turtles, probably inhabited restricted coastal marshes or brackish bays or estuaries (Thierstein and Berger, 1978; Gartner, 1979; Gartner and McGuirk, 1979; Thierstein, 1979). Kollman (1979) has found this hypothesis congenial to his investigations concerning distribution patterns and evolution of gastropods around the K–T boundary.

Brackish oceans and Permian extinctions

Extensive marine-derived salt deposits are known from Permian sequences in North America, the Soviet Union, western and central

Europe and South America. The idea that the progressive reduction in salinity of the world ocean during the latter half of Permian time led to massive extinction of marine organisms has been considered by a number of investigators (Fischer, 1964).

According to what Stevens (1977) considered to be a conservative estimate, about a tenth of the amount of salt present in modern oceans was extracted and stored in continental deposits during Late Permian time. That probably would have been sufficient to account for extinctions within families of organisms thought to be intolerant of significant changes in salinity, such as corals, crinoids, blastoids, ammonoids, and nautiloids. Life on land, plants as well as vertebrate animals, would not have been seriously affected by this crisis in the marine realm.

Changes in salinity related to Triassic extinctions

Holser (1977, 1984) has supported the idea that extinctions of Permian marine life were caused by a major drop in salinity of sea water, but he has proposed that even greater reductions in salinity took place during the Mesozoic. Early Triassic evaporites form a belt that originally must have stretched across Poland, East and West Germany, England and into Greenland.

To account for these salt deposits and for the extinctions that came after their accumulation, Holser proposed the following mechanism. Given a deep basin stretching along the margin of a continent, the sea water filling it would become progressively saltier through time under certain conditions. Suppose that the bottom of the basin is well below sea-level, but that between the basin and the open ocean a barrier rises to near the level of the sea. As the water over the basin evaporates, it would be replenished by waters flowing over the barrier. Thus over a period of a few hundred thousand years the water in the basin would become saltier and the surficial ocean water less so. Saline brine would sink to the bottom of the basin, where vast quantities of salts robbed from the sea would accumulate.

Meanwhile, shallow-water marine organisms adapted to waters of normal salinity would be subjected to increasing stress. With collapse of the barrier, there would have been catastrophic mixing of brine and ocean water. The quick return of the ocean to normal salinity would subject marine life to additional stress. Disruptions in marine ecology would not be restricted to near-surface environments. Some of the heavier brine would flow down to the bottom of the ocean, replacing colder with warmer water which initially was relatively deficient in oxygen.

Devonian extinctions attributed to incursion of anoxic water

The extinction horizon marking the Frasnian–Famennian transition at Medicine Lake, Alberta, is marked by a pyrite-rich siltstone ranging up to 5 cm thick. According to Geldsetzer *et al.* (1987), the pyrite formed during an incursion of oxygen-depleted water that was responsible for extinction of many species of shallow-water, bottom-dwelling organisms (see also Becker, 1986).

As for the source of the anoxic waters, the authors were uncertain. Possibly it discharged from an anoxic basin somewhere in the region. Or perhaps a cometary or asteroidal impact caused a turnover in an anoxic ocean, mixing oxygen-depleted water with oxygenated surface water. Whatever the cause, the authors did not propose that the event of extinction was synchronous on a global scale.

Devonian extinctions attributed to incursion of turbid water

To account for mass extinctions of marine organisms late in the Devonian, McLaren (1970) proposed that impact and explosion of a bolide in the ocean was a likely cause. He emphasized the selectivity of the extinctions. Apparently there was no sudden change in land plants and animals; and the types of marine organisms that disappeared at the close of the Frasnian were those that could not have survived in fresh water or in turbid water. Because he could conceive of no mechanism for spreading fresh water simultaneously over all the epicontinental seas and shelf margins of the world, he opted for turbid water. The turbulence generated by the giant waves and accompanying winds at the point where a large meteorite plunged into the ocean and exploded would induce a turbid oceanic environment that would persist far longer than bottom-dwelling, filter-feeding organisms and their larvae could endure.

Estimates of the time required for these extinctions have varied over a wide range. McLaren (1982) has proposed that the time might be measured in months rather than in years. Padian *et al.* (1984), who were inclined to favor the McLaren hypothesis, estimated a duration of something less than 500 000 years. McGhee, Jr (1982) considered that the extinction 'can correctly be called catastrophic', but not instantaneous, since the major portion of the event spread over at least 7 000 000 years.

Devonian extinctions attributed to cold-water oceans

Copper (1986) has suggested that cold-water spillover into equatorial regions attending major changes in paleogeography was a major factor in these extinctions. He cites evidence for Devonian glaciation in South America as indicative of stressful changes in global temperatures. (For discussion, see Raymond, *et al.*, 1987.)

Toxic oceans

Erickson and Dickson (1987) theorized that if a 10 km meteorite had hit the Earth and vaporized at the close of the Cretaceous, global dispersal of various components would have resulted in toxic levels of oceanic trace elements. Increased concentration of copper in the surficial waters, for example, would be injurious to phytoplankton, as nickel would be to diatoms, and as lead would be to nearly all forms of life. By the same token, an episode of intense volcanism might be attended by dispersal of other toxic trace elements such as cadmium and mercury.

10

Extinction of the dinosaurs

10.1 DINOSAURMANIA

In the United States at least, 1986 AD may be remembered as the year of the dinosaur. A front-page story in the October 21 edition of the *New York Times* carried the headline, 'It's Dinosaurs' Best Time in 70 Million Years'. The article detailed an amazing volume and variety of dinosaur-related items then flooding the market place with the approach of the Christmas season.

Articles of clothing imprinted with pictures of the giant reptiles included T-shirts, neckties, sweatsuits, scarves, pyjamas, and Halloween costumes. Earrings, buttons, coathooks, and toy banks were cast in the shapes of various dinosaurs. For children weary of rocking-horses, there was the rocking-dinosaur. A six-foot-tall inflatable model was available for the nursery. Nobody knows what colors adorned the skins of dinosaurs, but at almost any supermarket a child could find a dinosaur coloring-book, go home and put the imagination to work with crayons, and perhaps draw inspiration from reptilian reproductions on bedsheets or kites. Not all items were designed for children. Adult joggers and tennis-buffs could be fitted with T-shirts bearing fanciful images of Joggosaurus or Tennissaurus.

The causes underlying all this remain uncertain, though there has been speculation aplenty among amateur psychologists. Some say that dinosaurmania simply represents the market's search for novelty in promoting sales. Others suspect that the toy reptiles stimulate childish imagination by providing tangible substitutes for mysterious fairy-tale monsters such as dragons. Or perhaps these toys impart to the young a sense of superiority, as the big and fierce are transformed into the small and cuddlesome.

Psychology aside, this heightened interest in dinosaurs may reflect reports in the international press of some recent spectacular discoveries. What has been described as the largest cache of vertebrate fossils ever discovered in North America has turned up at a site on the Bay of Fundy in Nova Scotia. More than 100 000 pieces of bone, including skulls and jaws of reptiles thought to be the closest to mammals evolutionwise, were reported. What made this discovery especially interesting is the fact that the bones lie near the border between the Triassic and Jurassic systems. Thus they provide additional evidence for terminal Triassic mass extinctions (*Dallas Morning News*, Jan. 30, 1986).

Announcement in January of 1986 that remains of a huge herbivorous dinosaur had been found in the Late Jurassic rocks of New Mexico was still gaining press coverage seven months later (Mygatt, 1986). Judging by the dimensions of the bones, the animal was about 140 ft (42.67 m) long, stood 18 ft (5.49 m) high at the shoulder and 15 ft (4.58 m) high at the hips. It has been nicknamed 'seismosaurus' – the earth shaker – as taxonomic studies proceed (Figure 10.1).

The November 18, 1986 issue of the *New York Times* carried an announcement of the first discovery of dinosaur bones in Antarctica. Preliminary analysis indicates that this is probably a new species of

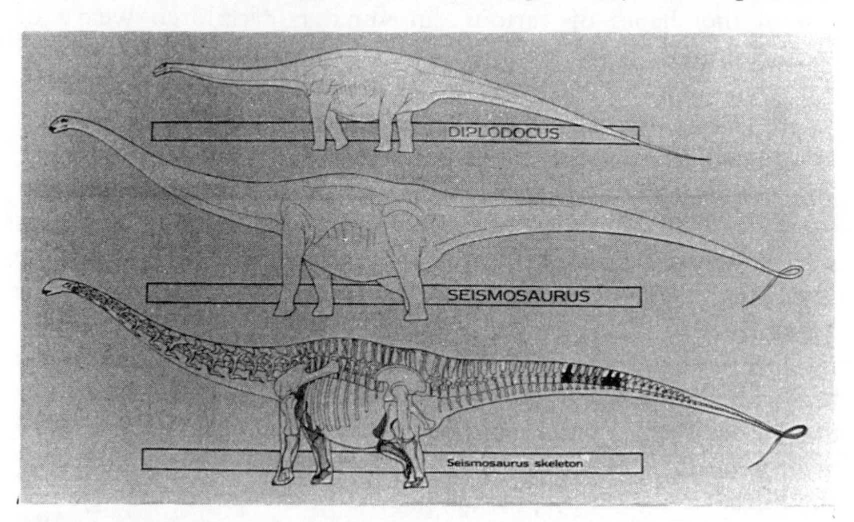

Figure 10.1 'Seismosaurus' compared in size with the giant sauropod *Diplodocus*, which with its length of 87.5 ft (26 m) was formerly thought to be the longest land animal that has ever lived. Artist's reconstruction by Douglas Henderson. Reproduced with permission of Dr David D. Gillette.

herbivore. Rocks in which the bones are embedded are dated at about 70 my, hence Maastrichtian in age.

Nine days later the *Daily Telegraph* announced that a dinosaur found in strata of Early Cretaceous age near Dorking had formally been named *Baryonyx Walkeri* (denoting 'a heavy-clawed creature found by Mr Walker'). Prior to this christening, during the time required for excavation, preparation and study, it had been nick-named 'Claws'. The reason: its claws measure 18 in (45.72 cm) long! Live weight was estimated at two tons.

Incidentally, paleontologists should be forever grateful to inquisitive and intelligent amateur collectors. *Baryonyx* was discovered by a plumber, 'seismosaurus' by a retired music teacher.

Mounting public interest in dinosaurs has led to the development of a new industry in the United States. Dinamation International of California was recently incorporated for the purpose of manufacturing life-sized models of these animals. Displays of their computerized robots, which move, bend, stretch, grunt, groan and roar have attracted crowds of visitors to museums around the country. Over a period of some seven months in 1987 and 1988, an exhibition of 14 of these monsters at a museum in Dallas, Texas, attracted 625 437 viewers. Adult admission: $5.00 a head.

10.2 DISCOVERY AND NAMING

The discovery of the terrestrial reptiles now called dinosaurs dates from 1822, when William Buckland described bones and teeth of a huge carnivorous Jurassic reptile which he named *Megalosaurus*. That same year Gideon Mantell, British physician and geologist, discovered remains of an Early Cretaceous herbivorous reptile which he named *Iguanadon*. In 1842 Richard Owen, foremost vertebrate paleontologist of his time, proposed Dinosauria ('terrible lizards') as the name to embrace most of the giant reptiles then described. H. G. Seeley, in 1877, subdivided the Dinosauria into two orders: the Saurischia, with pelves similar to those found in other reptiles; and the Ornithischia, with bird-like pelves.

10.3 POINTS OF GENERAL AGREEMENT CONCERNING DINOSAURS

What has been generally agreed as to the history of the dinosaurs is that they appeared in the latter third of the Triassic period; that both the herbivores and the carnivores evolved to occupy diverse habitats over most of the world; that they evolved into large sizes, none occupying niches available only to very small animals; that they did not become adapted to aquatic life, in the manner of crocodiles; and that one of their lines gave rise to birds (Olson and Thomas, 1980).

Numerous questions relating to the cause and precise time of their extinction remain unanswered. Of some 50 hypotheses that have been formulated to account for the demise of the dinosaurs, several have already been noted. Hypotheses profiled in the following section focus upon ecological changes in plant communities, alleged abnormalities related to reproduction, and thermal stress attending short- and long-term climatic change.

10.4 SOME HYPOTHESES FOR EXTINCTION

Harmful effects attending spread of flowering plants

The remarkable spread of flowering plants towards the close of the Cretaceous has occasionally been held accountable for the demise of the dinosaurs. These angiosperms were more efficient producers of oxygen than the plants they were replacing. Very large animals with very low rates of metabolism might become defunct when the concentration of atmospheric oxygen was increasing most rapidly. With greater concentrations of oxygen in the tissues of the dinosaurs, rates of metabolism would increase, so that it might have become impossible for these animals to consume enough food in order to satisfy bodies weighing tons. 'The dinosaurs may well have burnt themselves up, or out!' (Schatz, 1957; see also McAlester, 1970; Schopf *et al.*, 1971).

The adverse effects of too much oxygen was also cited as one of the possible causes of the decline and fall of the dinosaurs by Newell (1963). But he went on to speculate that the explosive evolution of pathogenic fungi that accompanied the rise of flowering plants may also have been a contributing factor.

Others have suggested that angiosperm foliage was bad for the digestion of herbivorous dinosaurs. Flowering plants produce toxic alkaloids, overdoses of which could probably be lethal. Moreover, with the spread of plants that seasonally lose their leaves, there would have been less food than required for the survival of the large herbivores, and when the herbivores died from poisoning, constipation or starvation, the meat-eating dinosaurs would starve to death.

Hypotheses such as the above have received little support in recent years. With the diversification of flowering plants over a period of about 40 my prior to the end of the Cretaceous, also came a simultaneous diversification of herbivorous dinosaurs (Russell, 1979). Also, ammonites, which joined the dinosaurs in extinction, didn't feed on angiosperms (Gould, 1985).

In a jocular vein, Dott (1983) has suggested that the dinosaurs sneezed to extinction. With the rise of flowering plants, the appearance of many pollens to which these animals had not previously been exposed may have produced violent allergic reactions 'that culminated in wholesale death by hay fever'. Mutations in the contemporary mammalian groups presumably had produced lineages resistant to bad effects of the new pollen. Therefore Dott proposed that the Late Mesozoic mammals might appropriately be called the 'flower children' of their times.

Bad eggs

Late Cretaceous dinosaur eggs from southern France and the Spanish Pyrenees presumably belonging to a species of the herbivore genus *Hypselosaurus* display two kinds of pathologic tendencies. In most cases the calcite shells are so abnormally thin that the embryos probably died of dehydration attending cracking of the shells. Less frequently, there is not one but two or more shells on single eggs, in which case the embryos probably died of suffocation.

By analogy with pathologic tendencies of the same kinds, reported for eggs of living birds and reptiles, these unusual dinosaur eggs have been attributed to hormonal imbalances in the bodies of the animals that laid them (Erben, Hoefs, and Wedepohl, 1979). What was here claimed as a possible cause contributing to the demise of a single species has not, however, been advocated as a cause for extinction of dinosaurs in general.

Thermal stress

Modern reptiles appear to exhibit greater tolerance toward temperatures much lower than those optimal for their activities than for temperatures only a little higher. For most strictly diurnal species of lizards in southern California, optimal temperature ranges from

37°C (98.6°F) to about 38°C (100.4°F). For snakes the range is between 31.6°C and 33°C. The lizards display discomfort if the temperature is raised only 3–4°C above the optimum, and die within the hour if the increase is a few degrees higher. On the other hand, lizards remain active with reduction of as much as 9°C below optimal levels. Both snakes and lizards may be maintained under refrigeration until their body temperatures drop to 1 or 2°C without suffering apparent harm (Cowles, 1939, 1940). Size and coloration appear to be factors in determining the levels of lethal thermal stress. Small body size is advantageous: the smaller the body the more rapid the cooling. Dark-colored lizards are more readily overheated than light-colored ones (LaMont, 1943).

Thermal stress has been repeatedly invoked as a factor in extinction of the dinosaurs. On the evidence of the Californian reptiles he examined, Cowles proposed that high temperatures may have been the chief factor in extinction. The large Cretaceous reptiles may have been driven from equatorial regions by high diurnal temperatures. Moving polewards, they were trapped by less intense but greatly prolonged exposure to lethal heat in savannas that offered little protection from the sun.

The brief thermal stress following impact of an asteroid or comet has been cited as a possible cause for demise of the dinosaurs (De Laubenfels, 1956; Hsü, 1982). Less brief but equally lethal climatic warming was the mechanism proposed by McLean (1978). Increased atmospheric carbon dioxide, following diminution of oxygen-producing phytoplankton during the Late Maastrichtian produced a greenhouse effect that caused an increase in global temperatures.

Others have proposed that the thermal stress was induced by changes from warmer to cooler environments. For example, Russell and Tucker (1971) reasoned that whereas a nearby supernova explosion may have heated upper levels of the atmosphere, any greenhouse effect would have been forestalled by disruption of the ozone layer and large-scale mixing of the atmosphere. Instead, high-altitude ice clouds would have reduced warmth from the sun. The resulting sudden shift toward cooler climates would have produced equally sudden environmental changes responsible for extinction of the dinosaurs.

That extinctions 'may have resulted primarily from stresses produced by decreasing equability of climate during the later Cretaceous' was the thesis of an investigation by Axelrod and Bailey

(1968). Draining of most epeiric seas led to a general decrease in equability over about the last 20 my of the Mesozoic. Most modern reptiles live in or adjacent to aquatic habitats in regions with mild climates, where annual departures from normal ranges of temperature are moderate. As for the dinosaurs, changes in mean annual temperatures of no more than a few degrees Fahrenheit, combined with increase in the mean annual *range* of temperatures could have imposed stresses to which they could not adapt. There could have been heat stress, cold stress, or both.

Hypotheses such as these raise questions as to how dinosaurs managed to control their body temperatures.

10.5 ENDOTHERMS OR ECTOTHERMS?

Until recently, all dinosaurs were presumed to have been 'cold-blooded' (ectothermal) like modern reptiles. Ectothermal animals require external sources of heat – whether derived directly from the sun, air, or ground – for elevating their body temperatures to optimal levels of activity. Until they have warmed to near optimal operating levels, they tend to be relatively inactive.

By contrast, 'warm-blooded' (endothermal) animals such as birds and mammals can adjust their body temperatures independently of external temperatures by internal metabolic processes. Beginning in the 1970s, the idea that dinosaurs must have been ectotherms simply because they are presently classified as reptiles has been challenged, and in the minds of some discredited (e.g. Desmond, 1976; Bakker, 1980, 1986).

Of the various arguments advanced for endothermy, those relating to posture and gait are counted among the more persuasive, even by the skeptics. Most dinosaurs were characterized by upright postures, as known from their trackways and as inferred from the articulation of their bones. Among living vertebrate animals, these characteristics are found only in endotherms, such as mammals and birds.

Upright posture implies that heads were above hearts – about 18 ft (5.49 m) above in the case of *Brachiosaurus*. Therefore blood-pressure must have been relatively high, probably powered by the double-pump action of the four-chambered heart found in modern mammals and birds.

Anatomy of legs, shoulders, and pelves in some bipedal dinosaurs suggests that they were adapted to running at speeds comparable to those of modern cursorial mammals and birds (Marx, 1978; Ostrom, 1980).

Another argument for endothermy is based on the global distribution of dinosaur remains. Late Cretaceous dinosaurs have been found along the north slope of Alaska in deposits of Late Campanian and Early Maastrichtian age. The paleolatitudes of these deposits have been estimated to range to 18° north of the Maastrichtian Arctic Circle. Presence of immature specimens in the collections suggests year-round residency. If so, these animals must have been able to endure weeks if not months of total darkness, and to survive reductions in temperature ranging down to 2–4°C for the coldest month of the year (Brouwers *et al.*, 1987; also see Gillette, 1986). However, that line of reasoning has been discounted by authors who maintain that Late Cretaceous climates were relatively warm on a global scale, and that in any case the animals could have migrated to warmer regions during winter-times (Marx, 1978). Oxygen isotope determinations have been cited as indicating that Late Mesozoic sea water was 10–20°C warmer than at present. Fossilized plant remains from a dinosaur site in Alaska suggest a comparison with climatic conditions presently prevailing at latitudes as low as 40–50°N. Evident redistribution of dinosaur remains by plate movements since Cretaceous time further reduces the weight of this zoogeographic argument (Ostrom, 1980).

Other arguments, pro and con, relate to the internal structure of bone, and to the ancestry of birds. The bones of certain dinosaurs preserve well developed hollow passages similar to the haversian canals responsible for conveying blood to bone cells of mammals and birds. But some mammals and birds have poorly developed haversian canals, whereas certain living reptiles, such as turtles, have such canals in some parts of their bones. Presumably, birds evolved from some dinosaurian lineage; and if so was the thermostat suddenly turned up at the instant of transition? But again, if there were warm-blooded dinosaurs, why did not some of them survive the end of the Mesozoic along with other endotherms? (For a review, see Wilford, 1985.)

The issue of endothermy remains unsettled. As Olson and Thomas (1980) have observed, 'We cannot put a thermometer in a dinosaur's cloaca'.

10.6 BANG OR WHIMPER?

Did the dinosaurian lineages just peter out, or did they suffer a sudden, catastrophic extinction? In either case, did the extinction occur before the end of the Cretaceous Period, at the end, or (heresy) afterward? Considering the spotty nature of the fossil record, can the experts arrive at a consensus on these issues?

The sequence of Late Cretaceous continental deposits exposed in Wyoming, Montana, and Alberta are world-famous for the abundance and diversity of the dinosaur remains they contain. The paleontological record, however, has been interpreted differently by different investigators.

According to Sloan *et al.* (1986), dinosaur extinction in the American West was a gradual process that began about 7 my before the end of the Cretaceous and accelerated rapidly during the final 0.3 my. Of 30 genera known to exist at the beginning of this decline, a maximum of only 12 survived until the end. These stratigraphically youngest specimens, mostly teeth, are found in sandstone-filled channels with a top 1.3 m above the coal bed ('lower Z coal') generally considered to mark the local K–T boundary. The authors estimate that this interval indicates that final extinction occurred about 40 000 years after the postulated asteroid produced the iridium anomaly reported for this region.

These claims have been challenged on various grounds by Retallack and Leahy (1986), Sheehan and Morse (1986), Bryant, Clemens, and Hutchison (1986) and by Retallack, Leahy, and Spoon (1987). A principal point at issue is whether isolated teeth found in the channels are contemporary with the sediments that enclose them, or whether they were reworked by streams from adjacent or underlying Late Cretaceous deposits. No articulated remains have been found in the Paleocene channels, whereas partial skeletons have been reported in undisputed channel fillings of Cretaceous age. The fact that many of the teeth show minimal evidence of abrasion could mean that, after reworking, they weren't transported very far. In fact, experimental tumbling of dinosaur teeth with water and sand indicated that the teeth were but little modified after rotations equivalent to 360–480 km of transport (Argast *et al.*, 1987).

Statistical analysis of data offered by Sloan *et al.* in support of their conclusion that extinction had been gradual led Sheehan and Morse to conclude that, on the contrary, the factual evidence suggests constant diversity and constant abundance of dinosaurs throughout the last 9 my of the Cretaceous. A similar view had been offered by Russell (1984). His tabulations of dinosaur genera indicated that 34 were living in North America during the Late Maastrichtian, compared with 27 for the Late Campanian–Early Maastrichtian.

That any apparent decline of dinosaurs toward the end of the Cretaceous may be a result of the relatively small sample of these animals recovered from the Maastrichtian as compared with older stages of the Cretaceous has been suggested repeatedly (e.g. Signor and Lipps, 1982; Russell, 1982; Clemens, 1982). Considering the world-wide distribution of dinosaurs during the Late Cretaceous, the total diversity of forms then in existence has probably been very poorly sampled, in which case the fossil record is not sufficiently well known to determine whether a long-term trend in diversity of dinosaurs did or did not occur prior to their extinction.

The Red Deer Valley of south-central Alberta has long been known as a prime collecting area for Late Cretaceous dinosaurs. There, as in the central Rocky Mountain foothills and in southwestern Saskatchewan, the boundary between the Cretaceous and Paleocene has been drawn at the bases of the oldest major coal seams, the lowest 1–2 cm of which show iridium peaks ranging between 1.35 ± 0.03 ppb and 5.6 ± 0.32 ppb. The boundary coincides with a shift from a diverse angiosperm flora of Late Maastrichtian age to a low-diversity Early Paleocene flora dominated by gymnosperm pollen and microspores. In the Red Deer drainage, the youngest dinosaur remains occur about 4 m below this break. Radiometric dates based on sanadine from seams of volcanic ash above and below the level of floral change give 63 ± 3 my as the best date for the contact between Cretaceous and Paleocene. According to Lerbekmo *et al.* (1979, 1987) these data confirm a regional coincidence between iridium anomalies, an extinction event in palynoflora, and initiation of coal formation. On the other hand, in western Canada there is an offset between the iridium peaks and the onset of a flora either dominated by or with a high percentage of fern microspores. The authors did not find the sum of evidence supportive of a simple North American-wide pattern of catastrophic destruction of vegetation coincident with an iridium event.

Russell (1979) was willing to consider the possibility that dinosaurs may have briefly survived the floral changes as reported at the K–T boundary in Alberta and elsewhere in the western interior of North America. Support for that speculation was later provided by Fassett (1982), who reported that dinosaur remains occur above that boundary, as defined on palynologic evidence in the San Juan Basin of New Mexico. More recent and more detailed stratigraphic studies there have provided additional support for Fassett's findings (Fassett, Lucas, and O'Neill, 1987).

According to Lindsay, Butler, and Johnson (1982), the youngest dinosaur remains in New Mexico occur in normally magnetized sediments, whereas the record of terminal extinction in Italy is in reversed polarity deposits below normally magnetized sediments. Therefore the extinction of marine biota in Italy must have happened well before the dinosaurs died out in New Mexico. That conclusion, and the evidence on which it was based, has been denied (Alvarez *et al.*, 1982).

Granting that the extinction of the dinosaurs in the San Juan Basin could indeed be correlated with the same reversal anomaly as reported from Gubbio, Padian and Clemens (1985) and Clemens (1986) have pointed out that this would not reveal whether the extinction of dinosaurs in the American West occurred precisely at the same time the microfossils were extinguished in Italy. That discordance in timing might have been 10 000, 100 000, or even 500 000 years. The data presently available, he proposed, support the view of a gradual transition from a dinosaur-dominated Cretaceous fauna to one characteristic of the Paleocene. With that proposition, Sullivan (1987) stands in agreement. His reassessment of sequences of reptilian taxa at the species level led him to conclude that the extinctions were gradual, and that there was no mass extinction of dinosaurs and other reptiles at the close of the Cretaceous.

Schopf (1982) considered the extinction of the dinosaurs at the end of the Cretaceous as no more than the conventionally accepted definition for the end of an era. He did't find 'anythng overwhelmingly noteworthy about the extinction of the most massive animals ever to have inhabited the lands of the Earth'. Granted that they were the largest terrestrial beasts, 'in a sense they were no different from the fastest or the slowest, or the fattest or the skinniest, or any other group of species which is no longer living'. In his view extinction is the normal way of life, considering that probably more than 99.99999% of all the species that have existed on Earth are now extinct.

So how did the world end at the close of the Cretaceous?, asked Van Valen and Sloan (1977). Not with a bang, not even with a whimper, they proposed, but with a slow plant succession which permitted the placental mammals to diversify and soon take over the Earth.

From the foregoing, one might conclude that dinosaurmania has affected the experts as well as the innocents. Reacting to the current confusion concerning the fate of the dinosaurs, Drake (in Silver, McLaren and Drake 1982) has suggested that 'the dinosaurs didn't become extinct, they evolved into red herrings'. On more serious taxonomic grounds, Bakker (1986) agrees that the dinosaurs have not become extinct. He would place dinosaurs in a class of their own, the Class Dinosauria, separate from and co-equal with the Class Reptilia. The birds he would include as a subclass (Aves) of the Dinosauria. Therefore, 'when the Canadian geese honk their way northward, we can say: "The dinosaurs are migrating, it must be spring!"'

11

Reactions to catastrophist hypotheses for mass extinctions

11.1 HYPOTHESES INVOKING IMPACTS OF EXTRATERRESTRIAL BODIES

There is a lingering suspicion among what appears to be a minority of investigators that the cryptoexplosion structures may be of volcanic rather than extraterrestrial origin (McCrea, 1981). But many who accept extraterrestrial origins have raised questions regarding the proposition that explosions attending impacts were responsible for mass extinctions. If any of the more violent impact scenarios is to be considered, then where are the accumulations of vertebrate and invertebrate remains at the Cretaceous–Tertiary boundary (Surlyk, 1980)? If the bolide plunged into the ocean, the giant waves there generated should have left sedimentary signatures in the form of turbidites. In the boundary sequences, intercalations of turbidites have indeed been reported at various horizons, but thus far never at the extinction levels (Smit, 1982; Van Valen, 1984). No crater or impact structure of the size and age required to cause the terminal Cretaceous extinctions has yet been identified.

As for the proposals that periodic comet showers were triggered by revolutions of a hypothetical Nemesis or Planet X, Sepkoski, Jr and Raup (1986) consider them not only to be highly speculative but also *ad hoc*, since the assumed periods of rotation and precession are based on the periods of cratering and extinction that are to be explained. Increased bombardment as the Solar System transits through spiral arms of the Milky Way Galaxy points to a periodicity in extinction events of about 50 my, almost twice the amount calculated on the basis of paleontological evidence.

An additional argument against these comet-shower hypotheses is based on analysis of iridium content in a piston core (Giant Piston Core 3) recovered from an abyssal clay section on the Pacific floor east of Hawaii (Kyte and Wasson, 1986). The core measured 24 m long. A 9 m segment appeared to provide a nearly continuous record of sedimentation over a span of about 34 my, ranging from 33 to 67 my ago. These figures would place the base in the Late Maastrichtian and the top in the Late Oligocene. Neutron-activation analyses of 149 samples of this segment showed a definite iridium peak only at the level taken as the boundary between the Cretaceous and Tertiary. There, the iridium content was 10 to 40 times greater than at higher levels. Even at this peak the concentration of that element is significantly less than current comet-shower models predict. The authors conclude that their findings cast 'serious doubts on the existence of periodicities in catastrophe-induced extinctions'.

11.2 HYPOTHESES INVOKING RADIATION

The proposition that bursts of radiation from nearby supernovae were responsible for extinctions has received little support in recent years. Granting that some of these bursts may have been large enough to be potential sources of biological effects, the means of detecting their arrival in the past are virtually non-existent (McCrea, 1981; Padian *et al.*, 1984).

Furthermore, the record provided by fossils offers little support for these hypotheses. The radiation should have affected terrestrial more than marine life, but there were times when most of the extinctions were in the sea. Land plants, which should have been the most exposed, were little affected during the terminal Permian and Cretaceous (Newell, 1963). With regard to the K–T transition, tropical plants with largely unprotected reproductive cells show much less change across the boundary than do temperate plants with more protected germ cells, and the most exposed intertidal and shallow-water subtidal invertebrates show the least change (Kauffman, 1984). In the case of all five major extinctions, benthic as well as planktonic organisms were sharply affected, thus ruling out hypothetical causes related to influxes of cosmic rays or high-energy radiation from supernovae or other astronomical sources (Jablonski, 1986b).

The suggestion that magnetic reversal 'spurts' are somehow

related to mass extinctions has also been questioned. With regard to the K–T transition, the reversal frequency began to rise about 10 my before the end of the Cretaceous period, and did not drop markedly until after a few million years into the Paleocene (Lutz, 1986).

11.3 PROBLEMS WITH IRIDIUM AND OTHER PLATINUM-GROUP METALS

Ordovician–Silurian boundary

The transitional section for these two systems at Anticosti Island, Quebec, has been cited as the most complete yet discovered in carbonate sequences containing abundant shells of marine invertebrates. Here the boundary is marked by a parting of calcareous silty clay 2–3 cm thick. Geochemical analyses of samples taken from about 2 m below the contact to the same distance above showed traces of iridium ranging from 5 to a maximum of 58 parts per trillion at the boundary. Throughout that interval, however, the concentrations of the trace element are simply proportional to the amount of aluminium-rich clay in the carbonate sequence. No spherules or shocked mineral grains were found. The clay at the boundary was probably transported to the sea by streams, according to Orth *et al.* (1986).

A second section of reference for this contact is exposed at Dob's Linn in the Southern Uplands of Scotland. There the boundary lies within a sequence of black, graptolite-rich shale interbedded with gray mudstone and thin beds of altered volcanic ash. Traces of iridium occur throughout the section, but there is no 'spike' at the boundary. The presence of the trace element has been attributed to detrital transport from host rocks formed deep in the Earth's crust or perhaps even in its mantle (Wilde *et al.*, 1986).

Late Devonian: Frasnian–Famennian boundary

An iridium anomaly of about twenty times above background has been reported from the Canning Basin of Western Australia. The anomaly was found in a stromatolite bed with a high content of iron oxide. In his commentary on that finding, McLaren (1985) pointed to the possibility that the extinction caused the anomaly. In the absence of extinguished species of browsing organisms, stromatolites may have flourished, and in fixing iron also concentrated platinum group metals.

That interpretation was also the one favored by McGhee, Jr *et al.* (1986). Extraction of trace elements from sea water by the bacteria reported to occur in the stromatolite bed they believed could account for the anomaly of only 300 parts per trillion of iridium. In any case they could find no such anomaly along the horizon marking the biological crisis in the Federal Republic of Germany, where the contact between the sequences in question is also lacking in shock-metamorphosed quartz and in spherules that could be interpreted as tektites.

Permian–Triassic boundary

A sequence of marine strata exposed in the area around Meishan, South China (about 200 km south-west of Shanghai) has been proposed as the world stratotype for the Late Permian. There the contact between the Permian and Triassic has been drawn along a bed of clay about 4 cm thick. Reports of anomalously high traces of iridium in the boundary clay have led to detailed investigations of the paleontology and geochemistry of strata in the contact zone. The results indicate that the clay contains only 0.002 ppb of iridium, as compared with ranges of 0.005–0.080 ppb for abundances reported in crustal rocks. Analyses of elements in the clay were interpreted to indicate that the parting probably originated as a silicic volcanic ash (Asaro *et al.*, 1982; Zhou, 1987).

Of six conodont species present in the Late Permian strata, at least four and possibly five survived into the Early Triassic. But there was a dramatic decrease in their numbers, from 322 specimens per kilogram of carbonate rock to 6 in the Early Triassic. This decrease persists for several meters above the boundary. If the boundary clay is indeed an altered ash, Clark *et al.* (1986) suggested that a massive volcanic eruption may have scattered enough tephra into the shallow seas of the Late Permian to cause the observed effects on marine organisms.

Triassic–Jurassic boundary

After reviewing the evidence for extinction of vertebrates across this boundary, Padian (1987) could find nothing in the pattern to suggest an abrupt extraterrestrial cause.

Cretaceous–Tertiary boundary

That concentrations of iridium occur along or near this boundary in both terrestrial and marine sedimentary deposits widely scattered over the globe has rarely been denied. However, many questions have been raised with regard to the source of this element and the processes responsible for its concentration.

Volcanic activity has been repeatedly invoked as an agency responsible for iridium anomalies. Until recently, the evidence has been indirect, based mainly on the composition of boundary clays that resemble altered volcanic ash. The first account of iridium's injection into the atmosphere by an active volcano came in 1983, following the eruption of Kiluea Volcano of Hawaii in January of that year. Airborne particles showed strikingly large concentrations of iridium; and the ratio of iridium to aluminum was found to be about 17 000 times the values obtained for Hawaiian lava.

That discovery led Zoller *et al.* (1983) to speculate that Hawaiian volcanos may be fed by magma ascending from great depths, possibly from the mantle. They suggested that major volcanic events, such as those responsible for the Deccan flood basalts of India, were of sufficient magnitude to have released the iridium found in boundary clays. Several other investigators have since agreed.

The Deccan volcanic rocks figure prominently in hypotheses that favor a terrestrial source for the irdium. Exposed over an area of about 500 000 km^2, they locally attain a thickness of more than 2100 m. On geochemical evidence the magma that formed these rocks has been identified with outpourings along a deep-seated rift in the ocean's floor. The minimal age for the sequence has been placed at 60 my (Chandrasekharam and Parthasarathy, 1978).

Duration of the Deccan event has not been firmly established. According to one estimate, the activity may have lasted less than a million years, in which case it probably ranks 'as one of the largest volcanic catastrophes in the last 200 my' (Courtillot *et al.*, 1986). A more recent critical examination of various ages that have been assigned to the volcanics, however, led Baksi (1987) to conclude that 'the current geochronological data do not permit close resolution of either the timing or the duration of Deccan volcanism. At present, the age of the bulk of the volcanics can only be placed in the range of about 70–60 my'.

Leaving aside the question of terrestrial or extraterrestrial origin, the question remains as to how the iridium came to be concentrated along or close to the K–T boundary. Various hypotheses call for mechanical concentration of particles by bottom currents, enriched residue from chemical destruction of host rock, or to geochemical processes operating in environments deficient in oxygen.

Kent (1981) has denied that a high concentration of iridium necessarily indicates an extraordinary extraterrestrial event. Even in marine sediments the boundary is marked by an hiatus, indicating an episode of non-deposition of terriginous sediments, or else one of erosion by bottom currents. In the first instance, iridium-bearing particles would settle down and accumulate in abnormally large amounts. In the second, currents would winnow the large iridium-rich spherules from the smaller clay particles. In either case, an anomalously high concentration of the trace element would result.

Rampino (1982) would account for the anomaly as the result of dissolution of normal pelagic limestone, producing a clay residue containing iridium-rich materials. A rise in the calcite compensation depth would have been the likely cause. He cited background levels for iridium in limestone above and below the contact at Gubbio of about 0.3 ppb on average. In the boundary clay, iridium averages about 6.4 ppb (maximum about 9.1 ppb), representing an enrichment factor of 20 to 30 times. Production of a 1 cm boundary clay would require the dissolution of only about 20 cm of the limestone. Carbonate dissolution has also been cited as a factor in producing a small iridium anomaly (0.4 ppb) at the boundary in Alabama (Donovan and Vail, 1986).

The claim has been made that platinum-group metals, such as iridium and osmium, may be concentrated by chemical processes affecting sediments that accumulate in environments deficient in oxygen. For many marine boundary clays, their richness in pyrite seems attributable to deposition under reducing conditions. In terrestrial situations, the platinum-group anomalies are commonly associated with coal (Kyte, Zhiming, and Wasson, 1980; Finkelman and Aruscavage, 1981; Rampino, 1982; Jablonski, 1986b).

The preferential association of iridium with manganese has been frequently noted. In 37 samples of deep-sea ooze gathered from the floors of the Caribbean and the Pacific–Antarctic Basin, the iridium content averaged 0.31 ± 0.14 ppb; and in all but one sample the iridium correlated strongly with the content of manganese (Crocket and Kuo, 1979). Concentrations of iridium equal to or greater than those reported from the boundary clays have been found in manganese nodules gathered from the floors of the Atlantic and Pacific oceans. For seven Pacific samples, the iridium content averaged 6.02 ± 1.10 ppb; for seven Atlantic samples the average was slightly less, 5.74 ± 0.22 ppb (Harriss *et al.*, 1968).

Claims that iridium anomalies occur precisely at the K–T boundary have been disputed. At Deep Sea Drilling Site 524 in the eastern South Atlantic, cores indicate that the anomaly occurs about 40 cm below the boundary, as indicated by microfossils. On the other hand, at DSDP Site 465 in the north-western Pacific the anomaly appears 30 cm above the biostratigraphic boundary. Based on estimated sedimentation rates, the extinction occurred about 80 000 years too late in the first instance, and about 100 000 years too early in the second (Jablonski, 1986b). Another discrepancy of this kind has been reported for the K–T sequence of Alabama, where the anomaly appears about 2 m above the contact (Reinhardt *et al.*, 1986).

11.4 EVIDENCE PROVIDED BY MICROSPHERULES

Microspherules, mainly of potassic feldspar, are more abundant within the boundary clay at Gubbio than in the beds above and below. But, according to Naslund, Officer and Johnson (1986) they also occur in that area in strata ranging in age from Turonian to Paleocene, representing a span of about 22 my. Therefore they are not unique to the boundary horizon. The authors submitted that the origin of the spherules does not seem related to a single extraterrestrial event at the close of the Cretaceous, nor to repeated impacts over a period of a million years or more. Instead, they proposed that the spherules probably formed by alteration of silicate glass droplets, perhaps reflecting a peak in volcanic activity.

Detailed examination of the K–T boundary interval at Caravaca led Izett (1987) to conclude that the potassium feldspar particles in the alleged 'impact layer' are not melt droplets formed by the impact of an extraterrestrial object, but were formed *in situ* after deposition of the enclosing sediment. Many of the feldspar particles are highly irregular in shape, and thus are not compatible with the hypothesis that they are melt droplets. Throughout the layer of iron-stained boundary clay the particles are poorly sorted according to size, suggesting that they did not settle through a water column.

Moreover, these particles are not confined to the boundary clay but also occur in stratigraphic units above and below it.

A deep-sea core from the Caribbean has been reported to contain a layer of sediment enriched in iridium and underlying a bed containing abundant microtektites. The layer coincides with the extinction of four or five species of Eocene radiolaria. Fission-track analyses of the microtektites indicate an age of 34.6 ± 0.6 my. That figure is close to the 34.2 my indicated for the age of widely distributed North American tektites on the basis of potassium-argon determinations. On that evidence Ganapathy (1982) concluded that a major meteorite impact had occurred during the Late Eocene.

More recent investigations indicate that there are at least three microtektite horizons in the oceanic sequence of Tertiary deposits: one in the early Late Eocene (38.5–39.5 my), another in the latest Eocene (37.5–38 my), and a third in the mid-Oligocene (31–32 my). None of these appears to coincide with mass faunal extinctions (Keller *et al.*, 1983; Jablonski, 1986b).

11.5 ARCTIC-SPILL HYPOTHESIS

As one line of evidence offered in support of this hypothesis, Gartner and Keany (1979) cited the occurrence of Danian coccolithophores alongside those of Late Maastrichtian age in cores taken from the bottom of the North Sea. Re-examination of the evidence led Perch-Nielsen *et al.* (1979) to conclude that this association was due to mechanical mixing of fossils of different ages in debris flows. Therefore there is no need to find an explanation for the assumed temporary intrusion of Danian assemblages into the North Sea (and nowhere else) during the Late Maastrichtian.

Benthonic organisms living below the approximately 30 m thick layer of brackish or fresh water should not have been affected by the Arctic spill, but many were; and since it does not seem reasonable that the Arctic waters could have diffused through the entire ocean column, some other explanation must be sought (McLean, 1981).

Kauffman (1984) has offered two arguments in opposition to the spill hypothesis, one mechanical and the other paleogeographic. The effluent fresh or brackish water issuing from the north polar region, he proposed, would have been trapped in the North Temperate and Pacific gyres, 'and thus would not spread to many parts of the world where extinctions nevertheless occurred'. Furthermore, the extinctions should have been greatest in shallow waters of the North Temperate Zone. But the record seems to show that extinction was most severe among tropical and subtropical biotas and decreased toward the poles.

Furthermore, geochemical studies of the latest Cretaceous sediments and of the fluids trapped within them provide no evidence of massive chemical changes in the ocean. Jablonski (1986b) has argued that latitudinal variations in patterns of extinction are not supportive of hypotheses invoking variations in ocean chemistry as primary causes of mass extinctions.

12

Alternative hypotheses for mass extinctions

12.1 RELATED TO VOLCANISM

Historical eruptions

Volcanism has often been proposed as a cause contributing to mass extinctions. Evidence derives in part from environmental changes documented for explosive eruptions in historic times, notably from the events at Mt Tambora, Indonesia; at Krakatoa; at Katmai in the Aleutions; and on Bali.

The Tambora eruption occurred on the island of Sumbawa on April 10 and 11 of 1815. Sounds of the explosions were heard at points as far distant as 2600 km. Prior to eruption, Mt Tambora rose about 4300 m above sea-level. Afterwards, a caldera 6 km across remained where the mountain had been. At its highest point, the rim of the caldera stood at an elevation of only 2850 m. Some 92 000 people died, directly or from starvation in the aftermath (Blong, 1984).

Tambora has been cited as the greatest known ash eruption of Holocene time. Ash fell at places as far distant as 1300 km. Its total volume has been estimated at 150 km^3 – twice the volume reckoned for the Krakatoa eruption. The dust veil penetrated the stratosphere; and over distances as great as 600 km pitch darkness prevailed for as long as two days. Mean temperatures for the Northern Hemisphere apparently dropped by 0.4° to 0.7°C. In any case, 1816 has been remembered as 'the year without a summmer'. Extremely cold summers were reported as far away as New England, and there were crop failures in Canada and Europe (Stothers, 1984).

Similar decreases in temperature have been correlated with three more recent major explosive eruptions. Following the Krakatoa explosion of 1883, the temperature in France remained about 10°C below normal for three years. After the explosion of Katmai in 1912, solar radiation in Algeria was reduced by about 20%. Following the 1960 eruption on Bali, a severe winter prevailed in the north-central and north-eastern United States (Axelrod, 1981).

Axelrod emphasized that environmental changes related to reduced temperatures were not the only causes for volcanically induced stresses on life. As examples of more immediately destructive consequences, he cited the devastation of coral reefs during the eruption of Krakatoa. Ashfall from the Vesuvian eruption of 1906 smothered benthonic life in waters nearby. A large lava flow issuing from Mauna Loa in 1950 entered the ocean, after which its underwater progress was marked by a lane of dead fish 60 m across.

Late Cretaceous volcanism

Observations such as these led Axelrod to conclude that episodic volcanism has been a significant factor in causing mass extinctions. As a prime example, he pointed to the increasing frequency with which layers of volcanic ejecta appear in the terrestrial and marine sedimentary rocks of Late Cretaceous age. During the last 10–15 my of Cretaceous time, episodes of explosive volcanic activity would quickly and repeatedly reduce global temperatures by 2° to 3°C, perhaps even more in high latitudes. A succession of pronounced cold waves might lead to the extinction of terrestrial plants and animals adapted to the 'stable conditions of a relatively isothermal world'. Choking on ash and dust, the dinosaurs and other large reptiles would have gradually been reduced in numbers and kinds. And if the plant foods needed by the large herbivores were destroyed for only one season they would starve, and so would the large carnivores thereafter.

Ashfalls in the ocean would decimate the plankters and disrupt the food chains dependent upon them. Ingestion of ash by marine invertebrates would lead to diverse digestive problems and ultimately to death.

McLean (1982) has been another proponent of volcanism as a factor in Late Cretaceous extinctions, which in his judgment likely spanned hundreds of thousands if not millions of years. Outpouring of Deccan lava from sources deep in the Earth would, among other matters, have brought vast quantities of carbon gases to the surface. Buildup of carbon dioxide in the atmosphere would have produced greenhouse conditions. As a consequence, organisms above a certain critical size (10 kg) would retain excessive bodily heat 'causing degeneration of internal germinal tissues in males, and, in females, disruption of calcium metabolism and hormonal systems'. Acidic volcanic gases injected into the oceans would cause dissolution of calcium carbonate. Iridium and osmium exuded with these gases would thereby be concentrated in the boundary clay layers.

With regard to the possible diverse effects on life attending Deccan volcanic activity, however, Toon (1984) has found the evidence unconvincing. Injection of large quantities of ash into the atmosphere would surely drive down temperatures in the short term. But according to his calculations, if all the explosive volcanos in the world were to erupt during a century, the mean temperature during that century would be only 1°C cooler than normal. Ash from each eruption would settle from the stratosphere within a few months. As for the ensuing greenhouse effect caused by injection of carbon dioxide into the atmosphere, Toon was equally skeptical. He assumed that Deccan volcanism produced 1 000 000 km^3 of lava at an average annual rate of 1 km^3. If so, then to have doubled the Earth's atmospheric carbon dioxide, more than a tenth of the Deccan lava would have had to be produced in about 100 years, which hardly seems reasonable. He concluded that the effects of volcanism on atmospheric conditions are probably insignificant compared with those attending impacts of large asteroids.

Keith (1982) and McLean (1987) have been among those holding that violent volcanism was mainly responsible for mass extinctions during the Late Cretaceous. Injection of large quantities of carbon dioxide into the atmosphere resulted in a greenhouse climate. Land animals suffered from thermal stress and exposure to such volcanic poisons as arsenic, fluorine and chlorine. Rising into the upper atmosphere, chlorine would deplete the ozone layer. Ultraviolet radiation would cause genetic damage, especially in animals like the dinosaurs that offered big targets.

Under the prevailing greenhouse conditions, any polar ice would have melted. Deep circulation of aerated polar waters would be retarded. Warm evaporite brines would sink to the deeps of what would become a virtually stagnant ocean. The final extinctions at the end of the Cretaceous may have resulted from several causes. Catastrophic mixing events at the stagnant culmination, such as those produced by hurricanes, tsunamis, and submarine volcanic eruptions, would have subjected marine organisms to stresses related to sudden changes in temperature and salinity of water. Widespread emergence of the continents would have deprived many organisms of the shallow-sea environment to which they were adapted.

As for the concentrations of iridium, Keith suggested that they originated as deposits in stagnant oceans, concentrated in sulfides and metal-organic compounds.

That release into the atmosphere of sulfur volatiles from fissure eruptions of basalt could have led to regional and possibly hemispheric darkness, cold weather and acid rain has been proposed by Rampino (1986). As an example, he cited the Roza flow of the Columbia River Basalt Group which was extruded about 14 my ago. Its volume has been estimated at 700 km^3. Based on reasonable assumptions, the aerosols released during that eruption were sufficient to induce global darkness of the kind envisioned in the more frightening scenarios for 'nuclear winters'.

Studies of boundary clays at the contact between the Cretaceous and Paleocene in Denmark, Italy, Spain, and Tunisia led Rampino and Reynolds (1983) to conclude that the mineralogy indicates a terrestrial source in glassy volcanic ash. The clays differ considerably in composition from place to place, but all the clay assemblages are typical of those found in limestone and mudrock within beds above and below the contact at local sections. Therefore volcanism must be considered as a possible explanation for the geochemical anomalies that have been reported.

Re-examination of the geochemistry across the K–T boundary around Gubbio has led Crocket *et al.* (1988) to opt for intense volcanic activity as the agency responsible for the iridium anomalies. They found that iridium is concentrated 63-fold over background in the boundary clay. On the other hand, the iridium enrichment is not confined to that clay parting, but extends about 2 m above and below it in the sections examined. Within this zone there are four iridium 'spikes' in addition to the one at the boundary. On that evidence the authors concluded that the processes responsible for these anomalies must have been sustained during a series of episodes over a period of some 300 000 years. If so, impact models calling for explosion of a single large bolide must be discarded.

Shock metamorphism attending volcanic explosions

Until recently, shock lamellae in mineral grains have been considered indicative of the violent explosive effects caused by impacts of extraterrestrial bodies. Studies of mineral grains in ash-flow deposits around the Toba caldera of Sumatra indicate that volcanic explosions can also account for shock metamorphism. In areal extent that caldera is more than 50 times larger than the one at Krakatoa. The latest eruption occurred about 75 000 years ago. Feldspar and mica crystals embedded in the ash-flow deposits show microstructures and textures indicative of shock. The biotite shows intensive kinking of lamellae, and planar features are present in the feldspar grains. Microstructures in quartz are rare, perhaps because most shock-lamellae 'healed' during the cooling of the superheated pyroclastics (Carter *et al.*, 1986).

Intense volcanism and selective extinctions during the terminal Cretaceous

The selectivity of extinctions at the end of the Cretaceous has been a persistent puzzle to theorists. If the great reptiles inhabiting land, sea and air disappeared, why then did so many types of snakes, mammals, fresh-water vertebrates, land plants, insects and marine invertebrates survive?

After reviewing the evidence at hand, Officer, Hallam, Drake, and Devine (1987) proposed that intense global volcanism, sustained over an interval of 10 000 to 100 000 years, and combined with pronounced regression of the sea lasting for a period of a few million years, may provide an answer to that question.

To begin, they cite findings to show that concentrations of iridium and associated elements, microspherules and shock deformation features are not necessarily diagnostic of impact by extraterrestrial bodies but can also be attributed to volcanic activity. Volatile emissions from this activity would lead to the precipitation of acid rain, reduce the alkalinity of the near-surface ocean waters, cause global changes in atmospheric temperature and deplete the ozone. 'These environmental effects coupled with those related to the major sea-level regression of the late Cretaceous provide the framework for an explanation of the selective nature of the observed extinction record'.

Present day emissions of sulfur dioxide, due to the combustion of fossil fuels, return to the ground or enter the ocean as acid rain. Studies of the impact of this polluted water on the Canadian flora and fauna indicate that some species are more tolerant than others to its ill effects. White pine and white birch are most sensitive to damage, whereas red cedar and sugar maple remain tolerant. In a Canadian lake that has undergone severe damage from acid rain, walleye and lake trout were the first to disappear, whereas yellow perch remain as survivors. If similar ranges in tolerance existed among species populating comparable habitats during the Late Cretaceous, selective extinctions should have been the result. Assuming that the emissions from the Deccan lavas were concentrated in a period of 10 000 years, then the magnitude of the volcanic acid rain would have been 14 times greater than at present. Clearly this could have had a significant and selective effect on the Cretaceous flora and fauna.

Acid rain falling into the ocean would reduce the alkalinity in near-surface waters in a chain of reactions by which sulfuric acid is reduced to carbonic acid. Carbon dioxide released from volcanic eruptions would also enter the ocean waters, further reducing their alkalinity. Again the extinctions were selective. For example, planktonic foraminifera which live in surficial waters, and which depend upon the availability of carbonate for their existence showed major extinctions. Conversely, deep water benthic foraminifera showed little change across the K–T boundary.

Global cooling of the Earth's surface temperature is known to have attended major volcanic eruptions of historical record. This cooling appears due not so much to injection into the atmosphere of dust – which falls out in a few months – as to the emission of aerosols, mainly of sulfur dioxide, which may remain in the stratosphere for one or two years. The authors estimated that Late Cretaceous volcanism may have resulted in global cooling in the order of 3°C. In that instance, selectivity would favor species adapted to cooler climates prevailing in relatively high latitudes.

Hydrogen chloride aerosols injected into the stratosphere during historic volcanic eruptions have led to a decrease in the protective ozone layer. By that token, increased volcanism during the Late Cretaceous would have caused a serious depletion in the ozone layer. Increase in exposure to ultraviolet radiation would selectively have favored animals with nocturnal habits and those living in subaquatic or subterranean habitats.

Flood basalt volcanism triggered by impact cratering

Rampino (1987) has suggested that flood-basalt eruptions, such as those responsible for the Deccan flows, could have resulted from large-body impacts. According to his calculations, a 10 km impactor should produce a crater 100–200 km across. The transient depth of the crater would be in the range of 20–40 km, deep enough to penetrate the 5 km of ocean crust, and even deep enough to excavate much or all of the thicker continental crust. Massive outpouring of basaltic lava would result, exacerbating the destructive effects attending the impact and explosion of the bolide.

12.2 RELATED TO CHANGES IN GLOBAL TEMPERATURE

On the premise that water temperature is the most important factor limiting geographic distribution of animal species in the ocean, Stanley (1984a, 1984b) has proposed that 'climatic cooling is the primary culprit behind most of the known marine crises'.

In some instances, direct evidence for global refrigeration has been found in ancient sedimentary deposits and contemporary land-forms attributable to episodes of continental glaciation. During the late Precambrian all continents, with the possible exception of Antarctica, were subjected to widespread glaciation. Crowell (1982) has noted the possibility of three glacial peaks at about 940, 770 and 615 millions of years ago. Concurrently, the ocean's populations of acritarchs (single-celled plankton) were decimated.

Likewise, the mass extinctions at the end of the Ordovician appear to coincide with episodes of continental glaciation, occurring over a period of about 50 my and lasting into the Early Silurian. Spotty glacial signatures occur in a belt extending from northern Europe into southern Africa, and from the Sahara region into Bolivia and Peru (Crowell, 1982; Hambrey, 1985; Barnes, 1986).

According to Crowell, the Late Paleozoic ice-age affected all continents. Beginning in the Late Carboniferous and ending in the Late Permian, ice caps waxed and waned over a period of about 90 my, ending about 5 my before the close of the Paleozoic.

Evidence for Late Devonian glaciation in what is now northern Brazil has been reported by Caputo (1985). According to his estimates, the glacial deposits spread over an area of 1 900 000 km^2 (see also Copper, 1977 and Kalvoda, 1986).

There appears to be no evidence for continental glaciation towards the end of the Cretaceous. The Late Cenozoic ice-age probably began locally during the older Tertiary with growth of the Antarctic ice cap. World-wide refrigeration appears to have set in by the Early Miocene, some 22 my ago.

As indirect evidence to support his hypothesis, Stanley called attention to geographic patterns of extinction in the paleontological record. In general, organisms adapted for life in warm tropical waters appear to have been the more vulnerable to climatic cooling. As an example, he cited regional differences in extinctions suffered by faunas during the present ice-age. Faunas of the western Atlantic and Caribbean were hardest hit, losing about 70% of their species. In the Mediterranean about 30% of the mollusk species disappeared. By contrast, the molluscan faunas of the Pacific coasts of California, Panama and Japan give no evidence of massive extinctions during glacial episodes. The reason: Pacific faunas could migrate north or south as water temperatures fluctuated, whereas the Mediterranean and Caribbean provided no easy escape routes.

Likewise, in the course of the Late Eocene extinctions, species inhabiting warm coastal waters suffered most. The same was true for the Late Cretaceous, when the belt of tropical oceans extending across south-eastern Asia, the Mediterranean and the Gulf of Mexico was the site of major extinctions. As for the Late Devonian extinctions, primitive corals and sponges were the most affected, whereas glass sponges, whose living representatives are mostly adapted to cool water, survived. That a decline in ocean temperatures was probably responsible for these extinctions was also the conclusion reached by Stearn (1987) after reviewing the evidence provided by stromatoporoids. Regarding the intervals of time required for mass extinctions to develop, Stanley thought these were 'geologically brief', amounting to no more than one or several million years.

The above views have been challenged on various grounds by Johnson (1984) and Colbath (1985). Since most of the pre-Cenozoic faunas that have been studied come from tropical or subtropical regions, there is a sampling problem. The highest stratigraphic level at which some species have been reported does not necessarily mean that the species in question became extinct at that level, but the assumption that it did will impart the impression of gradual extinction. During regressions of the sea, fossiliferous deposits are successively reduced in area available for future collecting, thus compounding the sampling error. Finally, of the extinction events recorded for the Paleozoic, only one (the Ordovician–Silurian) falls within the duration of a major glacial episode.

12.3 RELATED TO CHANGES IN SEA-LEVEL

Epeirogenesis and eustasy

In 1967 Newell proposed that the history of life on which our geologic time scale is based has been episodic rather than uniform. 'Modern paleontology', he predicted, 'will incorporate certain aspects of both catastrophism and uniformitarianism while rejecting others.' After reviewing the then current hypotheses to account for mass extinctions, he concluded that the most outstanding extinction episodes correspond to times of widespread emergence of the continents. Rapid emergence would result in catastrophic changes in both terrestrial and marine habitats, and such changes might well trigger mass extinctions. As for the length of time required for the extinctions, he set 6 my as a maximum, but suggested that these episodes may have lasted no longer than a few hundred or a few thousand years. Widespread emergences were attributed to vertical uplift of continents (epeirogenesis) and lowering of sea-level (eustasy).

Glacial, tectono-, and geoidal eustasy

Eustasy has been broadly defined to denote global or regional movements of the ocean surface in the vertical sense. Three major causes have been recognized. Sea-level falls as oceanic water is extracted during episodes of glaciation, and rises with the melting of the glaciers (glacial eustasy). As a result of earth movements, levels fall or rise whenever the volume of the ocean basins increases in the first instance or decreases in the second (tectono-eustasy). At present, global sea-level is characterized by highs and lows as much as 170 m in amplitude. These reflect differences in gravitation, in turn reflecting differences in mass at depths in the Earth beneath the ocean floors. Possibly the present configuration of these highs and lows reflects the pattern of humps, depressions, and eddies at the interface between the core and the mantle. If these buried masses are migrating, or if there should be changes in the speed of the Earth's rotation, the position and amplitude of the ocean's surficial highs and lows would also change (geoidal eustasy) (Mörner, 1984; Nunn, 1986).

Eustatic cycles

Eustasy has been conceived as a cyclic process. The interval of geological time during which a relative rise and fall of sea-level takes place on a global scale measures the magnitude of the cycle. Vail *et al.* (1977) proposed that there have been only two cycles of first magnitude, with durations between 200 and 300 my. The older commenced in the Late Precambrian and ended in Early Triassic time. The second began in the Middle Triassic and extends to the present. Superimposed on these first-order cycles are second- and third-order oscillations of sea-level, with durations between 1 and 80 my.

According to the above classification, the global swings in sea-level towards the end of the Mesozoic constitute a second-order eustatic cycle. Hays and Pitman (1973) mark the beginning of a rise near the boundary between the Lower and Upper Cretaceous (*c.* 100 my ago), and a crest at some time between the Turonian and early Maastrichtian (*c.* 90–70 my ago). Withdrawal began in the Late Maastrichtian and culminated in the Paleocene (*c.* 60 my ago). Due to the recent upsurge of interest in mass extinctions around the end of the Mesozoic, this particular cycle of about 40 my has been the subject of lively discussion.

Causes and results of changes in sea-level towards the end of the Mesozoic

Because there is no evidence for continental glaciation towards the close of the Mesozoic, glacial eustasy seems an unlikely cause for this cycle of sea-level change. Most investigators incline towards the view that the changes in level were due to changes in the cubic capacity of the ocean basins. These changes, in turn, have been attributed to changes in the volume or elevation of mid-ocean ridges attending seafloor spreading of crustal plates. During times of relatively rapid spreading, the cross-sectional area of the ridges would increase, the basins would become less capacious, and sea-level would rise. Mid-plate volcanism could produce a similar effect.

With reduction of spreading rate, or subduction of the ridges or subsidence of volcanic masses on the mid-plates, the capacity of the basins would increase and sea-level would fall (Vail *et al.*, 1977; Pitman, 1978).

On the other hand, Hallam (1987) proposes that the terminal Cretaceous regression was more likely due to vertical tectonic movements on the continents rather than to an increase in the cubic capacity of the ocean basins controlled by seafloor spreading.

Most estimates of the rise in sea-level at the Late Cretaceous maximum have fallen in the range of 150–350 m compared with the present level (Steckler, 1984). The encroaching seas must have covered more than a third of the continental surfaces. Beginning in the Late Cretaceous sea-level fell and, after a series of oscillations, assumed its present relatively low level. Stepwise emergence of the continents was attended by profound changes in climate. Global cooling in high latitudes and increased seasonal contrasts in temperature culminated in polar glaciation during the mid-Tertiary and mid-latitude glaciation in the Quaternary (Hays and Pitman, 1973).

Those who correlate mass extinctions with marked eustatic drops in sea-level point to a variety of environmental changes that would ensue. Seasonal extremes of temperature would adversely affect terrestrial animals. Marine organisms would suffer from more variable water temperatures and drastic reduction of habitable area (Hays and Pitman, 1973; Ager, 1981b; Hallam, 1984). Wiedmann (1986) also appealed to changes in sea-level as the cause for what he considered to have been the gradual decline of large invertebrates, such as ammonites and belemnites, towards the close of the Cretaceous. The turnover in calcareous oceanic plankton, however, he considered to have been a later 'instantaneous' event possibly related to impact of an extraterrestrial body (see also Kauffman, 1986).

Eustatic episodes during the Paleozoic era

Valentine and Moores (1970) have related major eustatic changes to assembly and fragmentation of continental masses. In the Late Precambrian, all the continents were evidently gathered as one (Pangaea). Fragmentation began during the Cambrian Period, followed by a reassembly to form Pangaea II in Permo–Triassic time. During the lengthy episode of fragmentation, growth of ridges at sites of seafloor spreading would displace water and cause sea-level to rise. Conversely, as continents are being joined, spreading stops, ridges subside, the volume of the ocean basin would increase and sea-level would fall.

With the formation of Pangaea II, increased seasonality of climate would affect terrestrial as well as marine life. On the continent there would be few barriers to migration; thus endemic biotas would be brought into association with one another. Extinctions and reductions in the diversity of life would attend the resulting competition among species.

The lengthy Paleozoic episode of continental fragmentation and reassembly cited by Valentine and Moores essentially coincides with the older first-order eustatic cycle identified by Vail *et al.* (1977). As noted previously, episodes of glacial eustasy towards the end of the Ordovician and near the terminal Paleozoic have been interpreted as events of a lesser order that none the less were factors in mass extinctions.

The boundary between the Mississippian and Pennsylvanian subsystems has not commonly been cited as marking a time of major extinctions. Recently, however, Saunders and Ramsbottom (1986) have called attention to the prominent unconformity separating these two units in North America, Europe, northern Africa, and elsewhere. Across that boundary 82% of ammonoid and 42% of crinoid genera became extinct. Conodonts and foraminifera also suffered a marked reduction in diversity. The authors attribute the unconformity to a eustatic event dating around 330 my ago. They were uncertain as to the cause, but suggested a possible link with onset of major volcanism in Australia dated at around 331 my ago.

Correlations between extinctions and eustatic episodes challenged

The alleged correlation between eustatic drops in sea-level and mass extinctions has been challenged on the ground that synchroneity has not been demonstrated except in the case of the terminal Cretaceous episode. The major drops in sea-level came before and after the end of the Permian, Berger (1984) has claimed. However that may be, Jablonski (1985) doubts that regression could account for extinction of the magnitude of the Permian–Triassic event, for the majority of marine families would persist in the undiminished shallow-water regions around oceanic islands. No major regressions of level appear to have been associated with the extinction episodes at the terminal Ordovician. Substantial and rapid fluctuations of sea-level during the Pleistocene did not result in widespread extinctions of marine shelf organisms, but paradoxically did affect organisms on land.

Staunch proponent of geoidal eustasy, Mörner (1981) has declared that global eustasy is but an illusion, and that it is fruitless to search for one global eustatic curve. Geoidal changes in sea-level related to changes in mass distribution attending continental drift and possibly in mantle convection negate the claim that erosional surfaces and unconformities in the marine record represent synchronous global regressions. Regional eustasy, he concludes, should be the new key phrase.

As Steckler (1984) has pointed out, there is wide divergence of opinion regarding the possibility of establishing a global eustatic curve. Although there is general agreement on long-term variations of sea-level over stretches of 200–300 my, on shorter time scales this agreement breaks down.

12.4 RELATED TO HISTORY OF PLANT LIFE

Major evolutionary developments in land plants

Three of the major extinction episodes have been correlated with evolutionary developments in land plants (Tappan, 1982, 1986). Initial radiation of vascular plants occurred during the Devonian, when the first large trees and forests appeared. In the Late Carboniferous, ferns and gymnosperm floras flourished. Radiation of the angiosperms followed in Cretaceous times. Each of these expansions would have affected the retention on land of organic carbon, together with associated nutrients such as phosphorus, which otherwise would have washed to sea. With a diminished supply of these essential nutrients, the phytoplankton populations would have been reduced, and the food-chain perturbations that followed would culminate in an episode of mass extinction within the marine realm.

This mechanism for extinction is slow-working. Vast amounts of organic carbon and associated nutrients were locked up in the coal beds from which the Carboniferous System got its name. But the ensuing gradual reduction in marine productivity did not result in the climactic Late Permian extinctions of marine life until several tens of millions of years thereafter.

At present more than half of the world's organic carbon is locked up in soil. With the diversification of the deciduous angiosperms beginning near the middle of the Cretaceous Period, soil first became in important carbon reservoir. As the Late Cretaceous seas withdrew from the continents, the deciduous flora and the soil cover associated with it progressively expanded and starved the adjacent oceans of nutrients for which much marine life had ultimately been dependent. Victims of this starvation included not only microscopic plankters but also, by food-chain reactions, the large secondary carnivores such as ammonites and marine reptiles.

As for the dinosaurs, they disappeared as their favored habitats shrank. Those reptiles were adapted to cycad–conifer savannas and scrublands, or else to fern and scouring-rush marshes. The spreading angiosperm forest canopies of the Late Cretaceous could neither harbor nor support them.

Tappan could find nothing in the paleobotanical record to support a 'lights-out' scenario for extinctions at the end of the Cretaceous. Cessation of photosynthesis during a blackout would have eliminated most tropical photosynthetic plankton in a matter of days, while a one-month blackout would have resulted in total collapse of food-chains.

Likewise Knoll (1984) could find no evidence for globally synchronous mass extinctions in the evolutionary record of vascular plants (ferns, fern allies and all seed plants). Granting that many families of 'seed ferns' and spore-bearing vascular plants disappeared towards the end of the Paleozoic, during times when conifers and other gymnosperm groups were expanding, these floral transitions were not synchronous on a global scale. Instead, they occurred on a continent-by-continent basis, probably over a period of about 25 my. As for extinctions in the Devonian and Late Cretaceous, the evidence is compelling that 'successive radiations of newly evolved families precipitated the decline and extinction of previously dominant groups'. Competition, perhaps accelerated by climatic changes, appears to have been the major determinant of large-scale changes presently recognized in the history of vascular plants.

Record of plant life in the western United States and Canada

The continental strata in these areas are rich in pollen and spores. Not uncommonly, thousands of these microscopic particles are found in a single gram of host rock. If the Cretaceous ended with a bang, some mark of that catastrophe should have been left on the palynologic record.

Prior to the discovery of the iridium anomaly in the American West, the palynologists had noted a floristic change across the boundary between the Cretaceous and Tertiary. For example, Hall and Norton (1967) found that in eastern Montana angiosperm species became less numerous, whereas the number of gymnosperm species increased. They suggested that a ‘modest deterioration of climate’ might have been responsible. From his studies of fossil floras in eastern Asia, Krassilov (1975) also concluded that the floristic change at the boundary records a cooling trend in the Late Maastrichtian.

Among the paleobotanists, reactions to hypotheses relating iridium anomalies to mass extinctions have been in the main skeptical. As Hickey (1984) observed, stratigraphical sections containing both the last dinosaurs and the latest Cretaceous plants have been carefully studied only in western North America. Even there, the Cretaceous flora persists above the level of the highest unreworked dinosaur bone: 6 m higher in Alberta, and at least 2–3 m higher in Wyoming. These gaps possibly represent 50 000–90 000 years. On a global scale, floral changes at the boundary were relatively minor and geographically variable. Granting that the record of the angiosperms provides no insight as to the causes of extinction, it is at least compatible with a model of climatic cooling (see also Worsley, 1971; Saito and van Donk, 1974).

A similar case for climatic changes has been made by Tschudy (1984). As for western America and Siberia, he granted that a case can be made for an increased rate of extinction of Late Cretaceous flora, but a massive elimination of taxa is not indicated. For each genus or species that disappeared at the boundary, several others persisted unchanged across it. Taken as a whole, the palynological record ‘supports the hypothesis of a gradually changing world flora across the boundary with no more abrupt changes than may be observed at epoch boundaries or even within subdivisions of epochs’. Therefore he found no reason to invoke supranormal or catastrophic mechanisms to account for the observed changes.

In Saskatchewan the boundary is marked by coincident anomalies in abundance of iridium and fern spores at the extinction level of a suite of Cretaceous pollen taxa. Locally, as many as 30% of the Cretaceous angiosperm species became extinct. The earliest Tertiary flora, however, is made up largely of surviving species that assumed new roles of dominance. Persistence of climatically sensitive taxa across the boundary indicates that if the paleoclimate was altered by some terminal Cretaceous event, it returned quickly to the pre-event condition (Nichols *et al.*, 1986). The authors were willing to consider the possibility that the perturbation in climate was a result of the hypothetical few months of darkness proposed by advocates of the impact scenario.

Recent detailed paleontological studies in the San Juan Basin of New Mexico indicate that Cretaceous pollen appears to be missing at or above the highest stratigraphic level at which dinosaur bone has been found. On the other hand, Paleocene pollen occurs at about the same level as the dinosaur remains. Thus if these remains have not been reworked, dinosaurs must have survived into the Paleocene, at least in this part of the world (Fassett, Lucas, and O'Neill, 1987). That possibility is strengthened by a recent report of dinosaur remains in Early Paleocene deposits of southern China (Sloan, 1987).

13 The new catastrophism

13.1 REVIVAL OF UNIFORMITARIANISM IN THE 20TH CENTURY

Radiometric dating of rocks proved that Kelvin's estimates of the Earth's age were far too short. Once all the time needed for a gradualistic approach to geological history had been restored, Lyellian uniformitarianism revived, and its popularity persisted in England and the United States well past mid-century.

When Leonard Hawkes, in 1957, delivered his address celebrating the 150th anniversary of The Geological Society of London, ages as great as 2200 my had been assigned to certain rocks on the basis of lead–uranium ratios. He ventured the following prediction, 'in the light of the discovery of the great span of geological time'.

> Fifty years ago it was reported that uniformitarians in the Lyellian sense were no more. On this 150th anniversary year I have to report that they are back, and in force. There are doubtless Fellows in this room who will be present at the 200th Anniversary of this Society. I venture to predict that they will then find Lyellian uniformitarianism to stand on an even broader and surer basis than it does today.

13.2 A CHANGE IN PERSPECTIVE

By a curious coincidence, Hawkes' prediction was published in the same year that the Soviet Union launched Sputnik I to usher in the space age. Thereafter geology exchanged an earth-bound for a cosmic perspective (e.g. Marvin, 1986; Kerr, 1987).

That new perspective was indelibly imprinted on our minds when first we viewed the image of the Earth as seen from the moon. In

1969, as the Apollo II astronauts spiraled from around the back of the moon, the Earth was seen to rise above the lifeless lunar landscape, much as we earthlings see the moon rise. From a distance of about 240 000 statute miles (386 232 km), our planet resembled a little blue and white marbled ball suspended in black space. Also one could not escape the thought that this object must be vulnerable to attack by any other objects that happened to be flying about.

At this writing, we are only 30 years into the space age, but the alterations in our views regarding time and change in the Solar System, and beyond, have been so enormous as to be counted revolutionary. Among other matters, increased attention is being paid to the possible roles of comets, asteroids, and meteorites in producing configurational changes on the Earth and other planets.

13.3 THE IMPACTORS

Meteorites

Among the identified flying objects (probably responsible for most unidentifiable flying objects, UFOs) the shooting stars or meteors are the ones most frequently observed. Bodies responsible for these displays (meteoroids) may fall to the ground as meteorites. Most meteorites are made primarily of silicate materials, but those made mainly of nickel and iron are easier to find and are the ones most frequently displayed in museums. Meteoroids may range in size from submicroscopic dust particles to bodies whose weights are measured in tons. The largest known meteorite, the Toba Iron, found in South West Africa, is some 9 ft (2.75 m) across and has a mass estimated at 66 tons (Wasson, 1985; Weaver, 1986).

Wasson has estimated that about 150 meteorites with masses greater than 500 grams (a little more than a pound) fall on land each year. Until recently, most of these objects were thought to be fragments of asteroids. That may still be the case, but a meteorite found in Antarctica appears to have been ejected from the moon. It is about the size of a golf ball. Fragments of feldspar-rich rock are set in brown glass; and in its mineralogy, chemical composition and texture it closely resembles specimens collected during the Apollo 16 mission to the lunar highlands (Marvin, 1986). Two additional lunar meteorites have since been recognized in collections from Antarctica. Evidently these were ejected by the impact and explosion of a

Figure 13.1 Meteorite believed to have been ejected from the Moon. Collected in 1981 on the ice sheet west of the Allan Hills, Antarctica. Cube is 1 cm across. (NASA photograph provided by

much larger body that probably produced a crater more than 3 km in diameter on the lunar surface (Melosh, 1985; Vickery and Melosh, 1987) (Figure 13.1).

Even more astonishing has been the discovery of eight meteorites whose physical and chemical properties strongly suggest that they originated on Mars. (See Gooding, 1987, for a discussion of problems relating to these objects.)

Asteroids

At present, about a thousand bodies of asteroidal appearance and with diameters greater than 1 km have been recognized as moving in orbits that intersect the orbit of the Earth. Weatherill and Shoemaker (1982) have estimated that three of these should impact Earth every million years. Impacts of objects 10 km in diameter – the size postulated by Alvarez and others to account for terminal Cretaceous extinctions – should occur about once every 40–50 my. As for bodies less than 1 km in diameter, calculations indicate that those measuring 0.5 km across should strike the Earth about 10 times per 1 my. Those striking land should produce craters around 10 km in diameter, or a little more than eight times the present diameter of Meteor Crater.

Comets

Because many astronomers consider comets to be primitive remnants of the dust and gas that formed the Solar System, the transit of Comet Halley in 1986 prompted intensive investigations through the launching of a fleet of spacecraft. Upon piercing the shroud of dust and gas, the spacecraft collected images of the nucleus which revealed that it is black, about 16 km long and 8 km wide. Ice makes up about 80% of the nucleus, so that the characterization of cometary nuclei as 'dirty snowballs' seems appropriate in this instance. Possible effects of cometary impacts on the Earth have been cited in Chapter 9. (For a summary account of findings and puzzles relating to the Halley transit, see Kerr, 1986a, 1986b.)

The possibility that comets far smaller than Halley may have left their marks on the moon has been raised by Donahue, Gombosi, and Sandel (1987). They propose that lunar craters with radii less than 1500 m may have been formed by impact and explosion of 'cometesimals', ice-covered boulders with radii ranging between a few meters and a few tens of meters. These bodies have not been observed directly; their presence is inferred from anomalous concentrations of hydrogen atoms between the orbits of the Earth and Mars, as detected by ultraviolet spectrometers on satellites.

13.4 IMPACT AS A FUNDAMENTAL PROCESS IN PLANETARY EVOLUTION

The fact that the visible surface of the moon is pitted by what appears to be craters was established in the 17th century by Galileo Galilei and Robert Hooke. In 1873 the British astronomer R. A. Proctor (1837–1888) proposed that these craters may have been caused by 'meteoric downfalls'.

Twenty years after the initiation of the space age, craters attributable to meteoritic impact had been photographed not only on the moon but also on the surfaces of Mercury, Venus and Mars, as well as on the Martian moons, all of which led Shoemaker (1977) to declare that 'the impact of solid bodies is the most fundamental of all processes that have taken place on the terrestrial planets'. (For an extended discussion of the mechanics of impact cratering, see Chapman and McKinnon, 1986.) (Figure 13.2.)

Strong support for that view was provided in the course of the Voyager 2 encounter with Uranus during late 1985 and early 1986. Ten satellites were discovered, bringing the total number of known satellites to 15. All showed evidence of impacts. Oberon, the outermost moon, is heavily cratered. Those nearer Uranus show signs of increasing disturbance related both to tectonic activity and to cratering. Miranda, innermost of the larger satellites, displays a patchwork surface suggesting that it has been disrupted by impact and reaccreted several times (Shoemaker, 1986; Waldrop, 1986).

13.5 COSMIC IMPACTS AND EXPLOSIONS

Astronomers theorize that at some time between 10 and 20 thousand million years ago the universe originated as a gigantic exploding fireball (van den Bergh, 1981). Following this 'big bang', matter accumulated to form galaxies. Each of these 'island universes' contains aggregates of stars, probably to be numbered in the thousands of millions. Until the 1970s each galaxy was generally

Figure 13.2 Photograph of the half-moon showing the abundance of lunar craters. (NASA photograph provided by Ursula Marvin.)

conceived to have evolved slowly, in isolation from its nearest neighbors. Recent advances in computer technology combined with observations from space vehicles have suggested a different and more violent sequence of events.

Galaxies have been classified according to their apparent overall shape. Some appear to be globular, ranging from the almost spherical to flattened ellipsoids; others, thinner, display spiral forms. According to a recent theory, the fatter galaxies may be 'star piles' resulting from collision and merger of two or more spirals. Cautious proponents of this theory speak of 'fierce galactic encounters' and of 'cannibalism' of smaller galaxies by super-giants. Thus collisions and mergers have been proposed as the dominant mechanisms underlying the evolution of the stellar universe (Schweizer, 1986).

Improved astronomical techniques have also disclosed that the galaxies display a distinctive structural pattern. They appear to trace out a network of 'bubbles' and voids, with the galaxies concentrated along the bubble walls – a sort of soapsuds structure (Waldrop, 1986). The scale is mind-boggling: apparent diameters of the bubbles range between 60 and 150 light years for that small slice of the universe thus far analyzed.

We seem to be moving towards a cosmogony featuring bangs, collisions and mergers as related evolutionary processes. If the Big Bang is to be taken as first-order in a hierarchy, then the presumed explosions that formed the cosmic bubbles would be second-order. Collisions and mergers of the galaxies would stand next in line. Recently a fourth-order bang and merger event has been proposed to account for the origin of our moon.

13.6 GIANT IMPACT THEORY OF LUNAR ORIGIN

In brief, this theory proposes that a body about the size of Mars struck the ancestral Earth nearly tangentially with a relative velocity around 10 km/s (Boss, 1986). At the time of collision, estimated at about 4 500 000 000 years ago, a rocky crust was beginning to form around the largely still molten protoearth. Attending the gigantic explosion upon impact, outer parts of both planets vaporized and jetted outward. The ejected materials derived both from the Earth and the impactor. A fraction of the ejected matter formed a disk-like band of vapors, liquids and particles circling the Earth. Matter in this 'prelunar disk' aggregated to form the moon. When the Earth recovered its globular shape, it combined parts of the impactor with its original substances. The entire process may have been completed in a hundred years or so.

Although this giant-impact scenario was put forward in 1975, it first became prominent in 1984, in the course of a conference on lunar and planetary science (Hartmann, Phillips and Taylor, 1986). Its proponents rely mainly on geochemical evidence. Moon rocks contain little or no iron, because they are formed from exterior portions of the two planets. They lack water, sodium and other volatile substances, which presumably boiled away during vaporization following impact. The Earth's outer mantle contains more of some heavy metals, such as gold and platinum, than geologists would expect: such metals should have sunk toward the core at the time when the primitive Earth was molten. Contamination of the Earth's shell would explain why deposits of these substances are shallow enough to be mined. As one enthusiastic advocate declared, 'The gold in your wedding ring came from the projectile, not the original earth' (Gleick, 1986).

In his review of the proceedings of the conference at which the giant impact theory became prominent, Stevenson (1986) offered the following observation: 'The recent revival of interest in lunar origin is remarkable since it does not arise from an infusion of new data but mainly from enthusiasm for some relatively new ideas that are still largely untested. Catastrophism is now in fashion. . . '. He went on to suggest that since the whole process may have taken only 100 years, it 'would have been fun to watch in real time'.

13.7 CATASTROPHIC CAUSES OF MASS EXTINCTIONS

At present there is no consensus among scientists regarding the cause or causes of mass extinctions. Most theorizing to date has centered around extinctions at or near the K–T boundary. The Alvarez hypothesis, which invokes an extraterrestrial agency, has sparked speculation among a diverse group of specialists including paleontologists, stratigraphers, sedimentologists, geochemists and geophysicists concerned with the dynamics of the solid Earth, meteorologists, climatologists, botanists, ecologists and astronomers. As indicated in the preceding chapters, there has been a negative reaction to each catastrophist hypothesis, reflecting in considerable measure the diverse viewpoints from which specialists in different disciplines weigh the evidence at hand. That bias was reflected in a recent opinion poll conducted by Hoffman and Nitecki (1985).

A random sampling of North American, British, German and Polish scientists taken in mid-1984 showed that only 31% of the 82 American geophysicists believed that impact of an extraterrestrial body was responsible for the K–T extinctions. Paleontologists were even more skeptical: only 9% of the 118 British respondents were believers, as compared with 14% of the 113 Germans and 16% of the 172 subscribers to the American journal *Paleobiology*.

The proposition that while there *was* an extraterrestrial impact at the K–T transition, other factors caused the mass extinction, drew the greatest support from *Paleobiology* subscribers (39%), British paleontologists (27%) and Polish geoscientists (40% of 122 respondents). That position is shared by Sepkoski, Jr (1986b), who observed that 'it remains possible that the end-Cretaceous impact was a rogue event that was coincidental with the operation of another periodic agent of extinction, amplifying its effects'. (See also Donovan, 1987.)

In answer to the question as to whether there was indeed an extraterrestrial impact at the boundary, 41% of the German paleontologists responded in the negative, as compared with like responses from 26% of Polish geoscientists and 23% of British paleontologists.

Two surprises (at least for this reader) emerged from the poll; 17% of German paleontologists and 13% of their British colleagues denied that there was *any* mass extinction at the K–T boundary. This viewpoint is in accord with a recent report on the K–T transition in Antarctica. There Zinsmeister *et al.* (1987) found that throughout an interval of 23 m above the contact, as defined by microfossils, ammonites are associated with typical Danian bivalves and gastropods. Second, this sampling of opinion revealed that the level of interest in the asteroid hypothesis has been distinctly higher in North America than in Europe. The pollsters suggested that this may be due to the popularization of the hypothesis in the mass media and in the weekly journal *Science*, combined with the American infatuation with dinosaurs.

A second poll of 118 vertebrate paleontologists on the question of a terminal Cretaceous impact was taken in the fall of 1985 (Raup, 1986a). Although 90% of the respondents did not deny the impact, only 4% accepted it as the major cause of the dinosaur extinctions;

and 27% felt that there was no mass extinction of land animals to be explained.

That mass extinctions may have been due to an extraordinary conjunction of causes rather than to the operations of single agencies is a proposition strongly defended by Gretener (1984). He illustrated his reasoning with an analogy drawn from a familiar game of chance. Given eight dice, probability theory predicts that the chance of throwing eight sixes in any one trial is about one in 2 000 000. But for 6 000 000 throws there is a 95% probability of eight sixes showing up at least once. Considering the Earth's antiquity, rare events such as the conjunction of planetoid impacts, increased volcanic emissions and eustatic lowering of sea-level cannot be ruled out as impossible. Following the lines of this argument, Lindinger and Keller (1987) suggested that the sharp and protracted fluctuations recorded in the stable-isotope stratigraphy across the K–T transition in Tunisia may have resulted from prolonged volcanic activity coupled with an extraterrestrial impact (See also Upchurch, 1987).

13.8 STATUS OF NEOCATASTROPHISM

The term 'neocatastrophism' is commonly associated with the speculations of Otto Schindewolf (1896–1971), the eminent German paleontologist. His paper entitled *Neokastrophismus*? was published in 1963. Schindewolf did not invent the term, and disliked being dubbed a neocatastrophist, for fear that he might be mistaken as one espousing the metaphysical catastrophism of earlier times. He made only two claims: mass extinctions are empirical facts, and these global phenomena permit the postulation of universal causes, such as episodic influxes of high-energy cosmic radiation. Appreciation for the significance of that argument has increased with the passage of time, and in 1977 Schindewolf's essay was reissued in English translation, appropriately in the journal *Catastrophist Geology*.

As applied to the ancient history of the Earth, catastrophism embodies two principal ideas. The first is that major changes in the Earth's configuration and in the successions of organic communities upon it have been episodic rather than gradualistic. Relatively brief episodes of rapid change interrupt longer intervals of more gradual changes. Or as Derek Ager, a self-confessed neocatastrophist, put it, 'The history of any one part of the earth, like the life of a soldier, consists of long periods of boredom and short periods of terror' (Ager, 1981a).

The second idea is that the historical record of configurational changes over a few thousand years is too brief to provide orders of magnitude for changes that have occurred during the past few thousand million years. For example we have not witnessed anything approaching the vast outpourings of lava during the Late Cretaceous, nor the catastrophic floods of the kinds attending the draining of glacial lakes in the northern Great Plains of the United States (Kehew and Lord, 1986). Fortunately we have not witnessed the impact and explosion of a meteorite capable of forming a hole the size of Meteor Crater. The Tungushka event was a frighteningly close encounter; and had that meteor exploded over London or New York City, neocatastrophism might have got off to a faster start than it did.

Thus the Lyellian maxim that 'the present is the key to the past' requires qualification. No one questions the proposition that the known natural laws respecting matter and energy must be assumed to have operated in the past. But Lyell's substantive uniformity, postulating that the rate of changes during the prehistoric past have on average been of about the same intensity as at present, is now recognized as an a priori assumption, just as Whewell and others claimed long ago. (For a statement of the various ways in which the present is atypical of the geological past, see Kauffman, 1986.)

Even so, substantive uniformity has its uses. When the catastrophist–uniformitarian debate was heating up in 1831, Sir John Herschel observed that:

> ...geologists have no longer recourse, as formerly to causes purely hypothetical...; but rather endeavor to confine themselves to a consideration of causes evidently in action at present, with a view to ascertaining how far they, in the first instance, are capable of accounting for the facts observed, and thus legitimately bringing into view, as residual phenomena, those effects which cannot be so accounted for.

Herschel's 'residual phenomena' are what we today call anomalies. In a geological context, it has been the discovery of anomalies that has called for the formulation of many radical new hypotheses to account for the evidence at hand. The evidence may be geochemical, as in the case of iridium anomalies; geomorphological, as in the case of craters manifestly formed by explosion but lacking evidence of associated volcanic activity; or structural, as in the case of cryptoexplosion features. Geomagnetic anomalies provide evidence for past lateral movements of continents. Gravimetric anomalies are the basis for the concept of geoidal eustasy; and geothermal anomalies guide the drill to new sources of energy as the Age of Petroleum approaches an end.

Thus the present appears to be a key to the inferential past, useful not only in curbing wild speculation regarding the Earth's ancient history, but also in identifying those anomalous situations that can't be accounted for in terms of human experience. In the latter sense, the past may be considered a key to the present.

In retrospect, the paleocatastrophists of the early 19th century seem not to have been so much the bad chaps in the black hats, at war with the white knights of the uniformitarian camp, as they were precursors of today's neocatastrophists. As Hooykaas (1970) has pointed out, metaphysical bias was present in the theories of some on both sides of the contest. But the old catastrophism bore a historical character, not only with regard to sequential and episodic changes in the Earth's configuration, but almost always with the idea of progress in the organic realm – the idea, as Hooykaas put it, that 'sudden geological outbursts run parallel with the rise of new (and also higher) animal types'.

13.9 A REVOLUTION IN THE EARTH AND PLANETARY SCIENCES?

Philosopher I. B. Cohen (1985) has identified four sequential tests to determine whether or not an intellectual movement qualifies as a revolution in science. First, there must be the testimony of contemporary witnesses to identify the times when the alleged revolution occurred. Later, there must be documentary evidence to show that in fact a radical restructuring of ideas and perspectives has occurred. Third, the proposition that a revolution has taken place must be affirmed by at least one card-carrying historian of science (the baptismal dip). Finally, the revolution must graduate into 'the mythology that is part of the accepted heritage of practicing scientists'. According to Cohen, the theory of continental drift has passed all four tests for being a revolution.

Perhaps we are in the midst of a second 20th century revolution involving the Earth and planetary sciences (Malone, 1985). Surely Shoemaker's bold assertion that 'the impact of solid bodies is the most fundamental of all processes that have taken place on the terrestrial planets' was a challenge to the conventional wisdom of the 1970s. The ideas that the Earth, like its sister planets, must have been repeatedly bombarded by extraterrestrial objects, and that some of these explosive episodes have altered the directions in which life otherwise would have evolved assuredly fulfill Cohen's criterion for a radical change in ideas and perspectives.

Revolution or not, many would agree with Dutch (1986) that these well may be 'the most intellectually stimulating times in the history of geology'. In the same vein, Picard (1985) has declared that 'working on a major problem is a lot better life than solving one. Life at the K–T boundary promises excitement for geologists for years to come'.

To date the controversies concerning mass extinctions and the role of extraterrestrial agencies in geohistory have been remarkably good-humored. That is not to deny that tart observations have occasionally turned up in the literature. Poking fun at the multitude of hypotheses formulated to account for the end-Cretaceous extinctions, Vogt and Holden (1979) noted that new data on that extinction 'have scarcely ruled out any past theory, but have fueled the promulgation of newer and even more outlandish proposals'. To the list of working hypotheses they proposed several others, including the possibility of a 'Late Cretaceous Noachian fleet of which only one ark survived'. Drake (Silver, McLaren and Drake, 1982) has called the extinction advocates 'paleonecrologists'. Prosh and McCracken (1985) have pointed out that the Alvarez hypothesis assumes a special significance from the viewpoint of the stratigrapher, inasmuch as an ideal geological boundary is provided by such a global and geologically instantaneous explosive event (see also Fassett and Rigby, 1987). They go on to point out that the imminent 'nuclear apocalypse' would also result in an excellent stratigraphic marker, in the form of a thin, highly radioactive and globally isochronous layer of sediment. In which case, they proposed that the Holocene should be renamed the 'kerocene' (from the Greek keros = death), or perhaps the 'Weshouldhave-scene'. 'Obscene' might be added to the list of possibilities.

13.10 PROGRESS OF THE EXTINCTION DEBATE

According to Thomson (1988) radically new scientific ideas follow a predictable course of nine stages of development before emerging as a new orthodoxy. Applied to the debate about terminal Cretaceous extinctions, he identified these stages as follows:

Recognition of an important unsolved problem

In this instance a mass extinction event involving dinosaurs.

Breakthrough

Discovery of the iridium spike at the K–T boundary: the Alvarez impact hypothesis.

Spreading the word

Involvement of the dinosaurs attracts widespread attention of the news media.

Excommunication of the apostates

Active proponents of the hypothesis and their converts become an exclusive club for a while.

Reaction

The radical new hypothesis is met with skepticism: by reasoned negative reactions from those who hold rival alternative views, and by knee-jerk negative reactions from those who oppose any radical new idea whatsoever.

Mobilization of resources

Members of diverse scientific disciplines join the debate.

Invention of new terminology, changing the terms of discussion

E.g. distinctions drawn between graded, stepwise, and catastrophic extinctions; background extinctions distinguished from mass extinctions; increasing concern with periodicities in extinction events.

The silly season

Marked by claims of earlier priority, proposal of ideas even more far-out than the original, and emphasis on extraterrestrial phenomena.

Calmer heads prevail as the subject becomes more complex

Both the original idea and its later manifestations are absorbed into a new mainstream. As Thomson notes, 'The impact theory is probably much more acceptable, for example, because it does not explain the extinction of the dinosaurs after all. Whatever the outcome of the debate', he concludes, 'nobody could ever say it was dull'.

13.11 A FAREWELL TO ISMS

Changes in geological and paleontological perspectives during the present century have been variously and collectively labeled neocatastrophism, actualistic catastrophism, and the new uniformitarianism. Each of these terms is ambiguous, and so should be abandoned.

As Simpson (1970) has pointed out, actualism has been used in two different senses. On the one hand, it stands for the proposition that past changes in terrestrial configurations may be accounted for by the operation of 'present' or 'now existing' causes, in which case it is generally synonymous with methodologic uniformitarianism. On the other hand, it has been identified as the postulate that the so-called natural laws have been and are unchanging, in which case it is generic to all historical sciences and not to historical geology alone.

As already indicated, the term catastrophism is encumbered with numerous ambiguities. Evidently Schindewolf was willing to accept neocatastrophism as a viable label for his views if, and only if, it were understood to be stripped of the supernaturalism, creationism, and providentialism formerly attached to it. Combining actualism with catastrophism to make actualistic catastrophism simply compounds the ambiguities resident in both isms. As for the 'new uniformitarianism', it remains a puzzle; if offered as a substitute for neocatastrophism, then we are left with the thankless task of sorting out the ideas attributable to each.

We don't need any more doctrinaire labels for recent advances in the natural sciences. Perhaps it is enough simply to claim that in the present ferment of ideas the principles of geohistory, biohistory and astrohistory are being subjected to critical re-examination toward the ambitious end of developing a unified theory for the unfolding of the universe.

References

Agassiz, L. (1840) *Etudes sur les Glaciers* (English translation and introduction by A.V. Carozzi), Hafner, New York and London, 1967.

Ager, D. (1981a) *The Nature of the Stratigraphic Record*, 2nd edn, John Wiley and Sons, New York.

Ager, D. (1981b) Major marine cycles in the Mesozoic, *J. Geol. Soc. Lond.* **138**, 159–66.

Alvarez, L.W. (1987) Mass extinctions caused by large bolide impacts. *Phys. Today*, **40**, 24–33.

Alvarez, L.W. *et al.* (1980) Extraterrestrial cause for the Cretaceous–Tertiary extinction. *Science*, **208**, 1095–108.

Alvarez, W. *et al.* (1982) Current status of the impact theory for the terminal Cretaceous extinction in *Geological implications of impacts of large asteroids and comets on the earth* (eds L.T. Silver and P.H. Schultz), *Geol. Soc. Am. Spec. Pap.*, **190**, pp. 305–15.

Alvarez, W. and Montanari, A. (1985) The sedimentary deposits of major impact events. *Geol. Soc. Am. Abst. with Programs*, pp. 512–13.

Alvarez, W. and Muller, R.A. (1984) Evidence from crater ages for periodic impacts on the Earth. *Nature*, **308**, 718–20.

Argast, S. *et al.* (1987) Transport-induced abrasion of fossil reptilian teeth: implications for the existence of Tertiary dinosaurs in the Hell Creek Formation, Montana. *Geology*, **15**, 927–30.

Argyle, E. (1986) Cretaceous extinctions and wildfires. *Science*, **234**, 261.

Armentano, T.V. and Woodwell, G.M. (1976) The production and standing crop of litter and humus in a forest exposed to chronic gamma radiation for twelve years. *Ecology*, **57**, 360–66.

Asaro, F. *et al.* (1982) Geochemical anomalies near the Eocene/Oligocene and Permian/Triassic boundaries in *Geological implic-*

ations of impacts of large asteroids and comets on the earth (eds L.T. Silver and P.H. Schultz), *Geol. Soc. Am. Spec. Pap.*, **190**, pp. 517–28.

Axelrod, D.I. (1981) Role of volcanism in climate and evolution. *Geol. Soc. Am. Spec. Pap.*, **185**.

Axelrod, D.I. and Bailey, H.P. (1968) Cretaceous dinosaur extinction. *Evolution*, **22**, 595–611.

Bakker, R.T. (1980) Dinosaur heresy–dinosaur renaissance in *A Cold Look at the Warm-blooded Dinosaurs* (eds E.C. Olson and R.D.K. Thomas), Westview Press, Boulder, Colorado, pp. 351–462.

Bakker, R.T. (1986) *The Dinosaur Heresies*, William Morrow and Co., New York.

Baksi, A.K. (1987) Critical evaluation of the age of the Deccan Traps, India: implications for flood-basalt volcanism and faunal extinctions. *Geology*, **15**, 147–50.

Barnes, C.R. (1986) The faunal extinction event near the Ordovician–Silurian boundary: a climatically-induced crisis in *Global bio-events* (ed O.H. Walliser), *Lect. Notes in Earth Sciences*, **8**, Springer-Verlag, Berlin, pp. 121–6.

Barringer, D.M. (1905) Coon Mt. and its crater. *Acad. Nat. Sciences, Philadelphia, Proc.*, **57**, 861–86.

Becker, R.T. (1986) Ammonoid evolution before, during and after the 'Kellwasser-event' – a review and preliminary new results in *Global bio-events* (ed O.H. Walliser), *Lect. Notes in Earth Sciences*, **8**, Springer-Verlag, Berlin, pp. 181–8.

de Beer, G., Sir (ed.) (1974) *Charles Darwin/Thomas Henry Huxley autobiographies.* Oxford University Press, London.

Berger, W.H., *rapporteur* (1984) Short-term changes affecting atmosphere, oceans, and sediments during the Phanerozoic in *Patterns of Change in Earth Evolution* (eds H.D. Holland and A.F. Trendall), Springer-Verlag, Berlin, pp. 171–205.

van den Bergh, S. (1981) Size and age of the universe. *Science*, **213**, 825–30.

Blong, R.J. (1984) *Volcanic Hazards*, Academic Press, Sydney.

Bohor, B.F. (1987) Dinosaurs, spherules, and the 'magic' layer: a new K–T boundary clay site in Wyoming. *Geology*, **15**, 896–9.

Bohor, B.F., Modreski, P.J. and Foord, E.E. (1987) Shocked quartz in the Cretaceous–Tertiary boundary clays: evidence for a global distribution. *Science*, **236**, 705–9.

Boon, J.D. and Albritton, C.C., Jr (1936) Meteorite craters and their possible relationship to 'cryptovolcanic structures'. *Field and Lab.*, **5**, 1–9.

Boon, J.D. and Albritton, C.C., Jr (1938) Established and supposed examples of meteoritic craters and structures. *Field and Lab.*, **6**, 44–56.

Boss, A.P. (1986) The origin of the moon. *Science*, **231**, 341–5.

Bourgeois, J. *et al.* (1988) A tsunami deposit at the Cretaceous–Tertiary boundary in Texas. *Science*, **241**, 567–70.

Branco, W. and Fraas, E. (1905) *Das Kryptovulkanische Becken von Steinheim*. K. Preuss. Akad. Wiss. Abh., pp. 1–64.

Brooks, R.R. *et al.* (1986) Stratigraphic occurrence of iridium anomalies at four Cretaceous/Tertiary boundary sites in New Zealand. *Geology*, **14**, 727–9.

Brouwers, E.M. *et al.* (1987) Dinosaurs on the North Slope, Alaska: high latitude, latest Cretaceous environments. *Science*, **237**, 1608–10.

Bryant, L.J., Clemens, W.A. and Hutchison, J.H. (1986) Cretaceous–Tertiary dinosaur extinction. *Science*, **234**, 1172.

Bucher, W.H. (1936) Cryptovolcanic structures in the United States. *Rep. 16th Internat. Geol. Congr.* US, **2**, pp. 1055–84.

Bucher, W.H. (1963) Cryptoexplosion structures caused from without or within the Earth ('astroblemes' or 'geoblemes'?). *Am. J. Sci.*, **261**, 597–649.

Buckland, W. (1823) *Reliquiae Diluvianae; or Observations on the Organic Remains Contained in Caves, Fissures, and Diluvial Gravel, and on Other Geological Phenomena Attending the Action of a Universal Deluge*. John Murray, London.

Buffon, G.L., Comte de (1962) *Des Époques de la Nature* (ed J. Roger), Musé. Natn. d'Hist. Naturelle, Mém., n.s., Sér C, **10**.

Burchfield, J.D. (1975) *Lord Kelvin and the Age of the Earth*. Science History Publications, New York.

Burnet, T. (1690–1691) *The Sacred Theory of the Earth*. Printed by R. Norton for Walter Kettilby, London, 2nd edn, 2 vols (reproduced, 1965, with Introduction by Basil Willey, by Southern Illinois University Press, Carbondale, Ill.).

Caputo, M.V. (1985) Late Devonian glaciation in South America. *Palaeogeog. Palaeoclimatol. Palaeoecol.*, **51**, 291–317.

Carozzi, A.V. (1984) Glaciology and the Ice Age. *J. Geol. Education*, **32**, 158–70.

Carter, N.L. *et al.* (1986) Dynamic deformation of volcanic ejecta from the Toba Caldera. Possible relevance to Cretaceous/Tertiary boundary phenomena. *Geology*, **14**, 380–7.

Cassidy, W.A. *et al.* (1965) Meteorites and craters of Campo del Cielo, Argentina. *Science*, **149**, 1055–64.

Chandrasekharam, D. and Parthasarathy, A. (1978) Geochemical and tectonic studies on the coastal and inland Deccan Trap volcanics and a model for the evolution of Deccan Trap volcanism. *Neues Jb. Miner. Abh.*, **132**, 214–29.

Chao, E.T.C., Shoemaker, E.M. and Madsen, B.M. (1960) First natural occurrence of coesite. *Science*, **132**, 220–22.

Chapman, C.R. and McKinnon, W.B. (1986) Cratering of planetary satellites in *Satellites* (eds J.A. Burns and M.S. Matthews), Univ. Arizona Press, Tucson, pp. 492–580.

Cisowski, S.M. and Fuller, M. (1986) Cretaceous extinctions and wildfires. *Science*, **234**, 261–2.

Clark, D.H., McCrea, W.H. and Stephenson, F.R. (1977) Frequency of nearby supernovae and climatic and biological catastrophes. *Nature*, **265**, 318–19.

Clark, D.L. *et al.* (1986) Conodont survival and low iridium abundances across the Permian–Triassic boundary in south China. *Science*, **233**, 984–6.

Clemens, W.A. (1982) Patterns of extinction and survival of the terrestrial biota during the Cretaceous/Tertiary transition in *Geological implications of impacts of large asteroids and comets on the Earth* (eds L.T. Silver and P.H. Schultz), Geol. Soc. Am. Spec. Pap., **190**, pp. 407–13.

Clemens, W.A. (1986) Evolution of the terrestrial vertebrate fauna during the Cretaceous/Tertiary transition in *Dynamics of Extinction* (ed D.K. Elliott), John Wiley and Sons, New York, pp. 63–85.

Cohen, I.B. (1985) *Revolution in Science*, Harvard University Press, Cambridge, Mass. and London.

Colbath, G.K. (1985) Comment on 'Temperature and biotic crises in the marine realm' by S.M. Stanley. *Geology*, **13**, 157.

Copper, P. (1977) Paleolatitudes in the Devonian of Brazil and the Frasnian–Famennian mass extinction. *Palaeogeog. Palaeoclimatol. Palaeoecol.*, **21**, 165–207.

Copper, P. (1986) Frasnian/Famennian mass extinction and cold-water oceans. *Geology*, **14**, 835–9.

Courtillot, V. *et al.* (1986) Deccan flood basalt at the Cretaceous–Tertiary boundary? *Geol. Soc. Am. Abst. with Programs*, p. 572.

Courtillot, V. and Besse, J. (1987) Magnetic field reversals, polar wander, and core-mantle coupling. *Science*, **237**, 1140–47.

Cowles, R.B. (1939) Possible implications of reptilian thermal tolerance. *Science*, **90**, 465–6.

Cowles, R.B. (1940) Additional implications of reptilian sensitivity to high temperatures. *Am. Naturalist*, **74**, 542–61.

Cox, A. (1969) Geomagnetic reversals. *Science*, **163**, 237–45.

Cox, L.R. (1942) New light on William Smith and his work. *Yorkshire Geol. Soc.*, **25**, 1–99.

Crocket, J.H. and Kuo, H.Y. (1979) Sources for gold, palladium and iridium in deep-sea sediments. *Geochem. et Cosmochem. Acta*, **43**, 831–42.

Crocket, J.H. *et al.* (1988) Distribution of noble elements across the Cretaceous/Tertiary boundary at Gubbio, Italy: Iridium variation as a constraint on the duration and nature of Cretaceous/Tertiary boundary events. *Geology*, **16**, 77–80.

Crowell, J.C. (1982) Continental glaciation through geologic time in *Climate in Earth History*, Studies in Geophysics, Natn. Acad. Press, Washington, D.C., pp. 77–82.

Cuvier, G. (1817) *Essay on the Theory of the Earth, with Mineralogical Notes and an Account of Cuvier's Geological Discoveries by Professor Jameson*. (Translation of *Discours sur les Révolutions de la Surface du Globe*), 3rd edn, W. Blackwood, Edinburgh.

Darwin, C. (1859) *On the Origin of Species by Natural Selection*. John Murray, London.

Davis, M., Hut, P. and Muller, R.A. (1984) Extinction of species by periodic comet showers. *Nature*, **308**, 715–17.

De Laubenfels, M.W. (1956) Dinosaur extinction: one more hypothesis. *J. Paleont.*, **30**, 207–18.

De Maillet, B. (1968) *Telliamed, or the Conversations between an Indian Philosopher and a French Missionary on the Diminution of the Sea* (ed and trans., A.V. Carozzi), Univ. of Illinois Press, Urbana.

Dence, M.R., Grieve, R.A.F. and Robertson, P.B. (1977) Terrestrial impact structures: principal characteristics and energy considerations in *Impact and Explosion Cratering* (eds D.J. Roddy, R.O. Pepin and R.B. Merrill), Pergamon Press, New York, pp. 247–75.

Dence, M.R. and Guy-Bray, J.V. (1972) The Brent Crater in *Some*

Astroblemes, Craters, and Cryptovolcanic Structures in Ontario and Quebec; Guide to Excursion A65, 24th Internat. Geol. Congr., Montreal, Quebec, pp. 11–18.

Desmond, A.J. (1976) *The Hot-blooded Dinosaurs: a Revolution in Paleontology*, Dial Press, New York.

Dietz, R.S. (1961) Vredefort ring structure: meteorite impact scar? *J. Geol.*, **69**, 499–516.

Dietz, R.S. (1963) Cryptoexplosion structures: a discussion. *Am. J. Sci.*, **261**, 650–64.

Donahue, T.M., Gombosi, T.I. and Sandel, B.R. (1987) Cometesimals in the inner solar system. *Nature*, **330**, 548–50.

Donovan, A.D. and Vail, P.R. (1986) Sequence stratigraphy of the K–T boundary in Alabama: a noncatastrophic alternative. *Geol. Soc. Am. Abst. with Programs*, p. 587.

Donovan, S.K. (1987) Mass extinctions. How sudden is sudden? *Nature*, **328**, 109.

Dott, R.H., Jr (1983) Itching eyes and dinosaur demise. *Geology*, **11**, 126.

Dutch, S. (1986) Rare events discussed. *Geotimes*, **31**, 4.

Emiliani, C., Kraus, E.B. and Shoemaker, E.M. (1981) Sudden death at the end of the Mesozoic. *Earth Planet. Sci. Letters*, **55**, 317–34.

Erben, H.K., Hoefs, J. and Wedepohl, K.H. (1979) Paleobiological and isotopic studies of eggshells from a declining dinosaur species. *Paleobiology*, **5**, 380–414.

Erickson, D.J., III and Dickson, S.M. (1987) Global trace-element biochemistry at the K/T boundary: oceanic and biotic response to a hypothetical meteorite impact. *Geology*, **15**, 1014–17.

Farsan, N.M. (1986) Frasnian mass extinction – a single catastrophic event or cumulative? in *Global Bio-events* (ed O.H. Walliser), *Lect. Notes in Earth Sci.*, **8**, Springer-Verlag, Berlin, pp. 189–97.

Fassett, J.E. (1982) Dinosaurs in the San Juan Basin, New Mexico, may have survived the event that resulted in creation of an iridium-enriched zone near the Cretaceous/Tertiary boundary in *Geological implications of impacts of large asteroids and comets on the Earth* (eds L.T. Silver and P.H. Schultz), Geol. Soc. Am. Spec. Pap., **190**, pp. 435–47.

Fassett, J.E. and Rigby, J.K., Jr (eds) (1987) *The Cretaceous–Tertiary boundary in the San Juan and Raton Basins, New Mexico and Colorado*, Geol. Soc. Am. Spec. Pap., **209**.

Fassett, J.E., Lucas, S.G. and O'Neill (1987) Dinosaurs, pollen and spores, and the age of the Ojo Alamo Sandstone, San Juan Basin,

New Mexico in *The Cretaceous–Tertiary Boundary in the San Juan and Raton Basins, New Mexico and Colorado* (eds J.E. Fassett and J.K. Rigby, Jr), Geol. Soc. Am. Spec. Pap., **209**, pp. 17–34.

Finkelman, R.B. and Aruscavage, P.J. (1981) Concentration of some platinum-group metals in coal. *Internat. J. Coal Geol.*, **1**, 95–9.

Fischer, A.G. (1964) Brackish ocean as the cause of the Permo–Triassic marine faunal crisis in *Problems in Paleoclimatology* (ed A.E.M. Nairn), Intersciences Publishers, London, pp. 566–77.

Fischer, A.G. (1984) The two Phanerozoic supercycles in *Catastrophes and Earth History* (eds W.A. Berggren and J.A. van Couvering), Princeton University Press, Princeton, New Jersey, pp. 129–50.

Fischer, A.G. and Arthur, M.A. (1977) Secular variations in the pelagic realm in *Deep-water Carbonate Environments* (eds H.E. Cook and P. Enos), Soc. Econ. Paleontologists and Mineralogists, Spec. Pub. 25, pp. 19–50.

Flessa, K.W., *rapporteur* (1986) Causes and consequences of extinction in *Patterns and Processes in the History of Life* (eds D.M. Raup and D. Jablonski), Springer-Verlag, Berlin, pp. 235–57.

Ganapathy, R. (1980) A major meteorite impact on the Earth 65 million years ago. *Science*, **209**, 921–3.

Ganapathy, R. (1982) Evidence for a major meteorite impact on the Earth 34 million years ago: implications on the origin of North American tektites and Eocene extinction in *Geological implications of impacts of large asteroids and comets on the Earth* (eds L.T. Silver and P.H. Schultz), Geol. Soc. Am. Spec. Pap., **190**, pp. 513–16.

Ganapathy, R. (1983) The Tungushka explosion of 1908: discovery of meteoritic debris near the explosion site and at the South Pole. *Science*, **220**, 1158–61.

Ganapathy, R. *et al.* (1981) Iridium anomaly at the Cretaceous–Tertiary boundary in Texas. *Earth Planet. Sci. Letters*, **54**, 393–6.

Gartner, S. (1979) Terminal Cretaceous extinctions: a comprehensive mechanism in *Cretaceous–Tertiary Boundary Events* (ed W.K. Christensen and T. Birkelund), **2**, University of Copenhagen, Copenhagen, pp. 26–8.

Gartner, S. and Keany, J. (1979) The coccolith succession across the Cretaceous–Tertiary boundary in the subsurface of the North Sea in *Cretaceous–Tertiary Boundary Events* (eds W.K. Christensen and T. Birkelund), **2**, University of Copenhagen, Copenhagen, pp. 103–5.

Gartner, S. and McGuirk, J.P. (1979) Terminal Cretaceous extinctions: scenario for a catastrophe. *Science*, **206**, 1272–6.

Geldsetzer, H.J. *et al.* (1987) Sulfur-isotope anomaly associated with the Frasnian–Famennian extinction, Medicine Lake, Alberta, Canada. *Geology*, **15**, 393–6.

Gerstl. S.A.W. and Zardecki, A. (1982) Reduction of photosynthetically active radiation under extreme stratospheric aerosol loads in *Geological implications of impacts of large asteroids and comets on the Earth* (eds L.T. Silver and P.H. Schultz), Geol. Soc. Am. Spec. Pap., **190**, pp. 201–10.

Gilbert, G.K. (1896) The origin of hypotheses, illustrated by the discussion of a topographic problem. *Science*, **3**, 1–13.

Gillette, J.L. (1986) Dinosaurs found at North and South poles. New Mexico Mus. Nat. Hist., Notes from the Underground, **I**, 3.

Glass, B.P. (1969) Silicate spherules from Tungushka impact area. *Science*, **164**, 547–9.

Glass, B.P. (1982) Possible correlations between tektite events and climatic changes in *Geological implications of impacts of large asteroids and comets on the Earth* (eds L.T. Silver and P.H. Schultz), Geol. Soc. Am. Spec. Pap., **190**, pp. 251–6.

Gleick, J. (1986) Moon's creation now attributed to a giant crash. *New York Times*, June 3, pp. 19, 21.

Gooding, J.L. (1987) Are SNCs smoked or salted? *Geotimes*, **32**(6), 10–11.

Gostin, V.A. *et al.* (1986) Impact ejecta horizon within the Late Precambrian shales, Adelaide geosyncline, South Australia. *Science*, **233**, 198–200.

Gould, S.J. (1965) Is uniformitarianism necessary? *Am. J. Sci.*, **263**, 223–8.

Gould, S.J. (1985) Sex, drugs, disasters, and the extinction of the dinosaurs in *The Flamingo's Smile*, W.W. Norton and Co., New York, London, pp. 417–26.

Gretener, P.E. (1984) Reflections on the 'rare event' and related concepts in geology in *Catastrophes and Earth History* (eds W.A. Berggren and J.A. van Couvering), Princeton University Press, Princeton, New Jersey, pp. 77–89.

Grieve, R.A.F. (1982) The record of impact on Earth: implications for a major Cretaceous/Tertiary impact event in *Geological implications of impacts of large asteroids and comets on the Earth* (eds L.T. Silver and P.H. Schultz), Geol. Soc. Am. Spec. Pap., **190**, pp. 25–37.

Hall, J., Sir (1805) Experiments on whinstone and lava. *R. Soc. Edinburgh, Trans.*, **5**, 43–98.

Hall, J.W. and Norton, N.J. (1967) Palynological evidence for floristic change across the Cretaceous–Tertiary boundary in eastern Montana (USA). *Palaeogeog. Palaeoclimatol. Palaeoecol.*, **3**, 121–31.

Hallam, A. (1984) Pre-Quaternary sea-level changes. *Earth Planet Sci. Ann. Rev.*, **12**, 205–43.

Hallam, A. (1987) End-Cretaceous mass extinction event: argument for terrestrial causation. *Science*, **238**, 1237–42.

Hambrey, M.J. (1985) The Late Ordovician–Early Silurian glacial period. *Palaeogeog. Palaeoclimatol. Palaeoecol.*, **51**, 273–89.

Hansen, T.A. (1984) Sedimentology and extinction patterns across the Cretaceous–Tertiary boundary in East Texas in *The Cretaceous–Tertiary Boundary and Lower Tertiary of the Brazos River Valley* (ed T.E. Yancey), South Texas Geol. Soc., San Antonio, pp. 21–36.

Harriss, R.C. *et al.* (1968) Palladium, iridium, and gold in deep-sea manganese nodules. *Geochem. et Cosmochem. Acta*, **32**, 1049–56.

Hartmann, W.K., Phillips, R.J. and Taylor, G.J. (eds) (1986) *Origin of the Moon*. Lunar and Planetary Inst., Houston, Texas.

Hawkes, L. (1957) Some aspects of the progress in geology in the last fifty years, *I. Geol. Soc. London, Q.J.*, **113**, 309–21.

Hays, J.D. and Pitman, W.C., III (1973) Lithospheric plate motion, sea-level changes and climate and ecological consequences. *Nature*, **246**, 18–22.

Herschel, J.F.W., Sir (1831) A preliminary discourse on the study of natural philosophy in *The Cabinet of Natural Philosophy conducted by the Rev. Dionysius Lardner*, Carey and Lea, Philadelphia.

Hickey, L.J. (1984) Changes in angiosperm flora across the Cretaceous–Tertiary boundary in *Catastrophes and Earth History* (eds W.A. Berggren and J.A. van Couvering), Princeton University Press, Princeton, New Jersey, pp. 279–313.

Hoffman, A. and Nitecki, M.H. (1985) Reception of asteroid hypothesis of terminal Cretaceous extinctions. *Geology*, **13**, 884–7.

Holmes, A. (1937) *The Age of the Earth*. Nelson, London.

Holser, W.T. (1977) Catastrophic chemical events in the history of the ocean. *Nature*, **267**, 403–408.

Holser, W.T. (1984) Gradual and abrupt shifts in ocean chemistry in *Patterns of Change in Earth Evolution* (eds H.D. Holland and A.F. Trendall), Springer-Verlag, Berlin, pp. 123–43.

Hooke, R. (1705) Lectures and discourses of earthquakes and subterraneous eruptions . . . in *The Posthumous Works of Dr. Robert Hooke* (ed R. Waller), Smith and Walford, London, pp. 279–450.

Hooykaas, R. (1970) Catastrophism in geology, its scientific character in relation to actualism and uniformitarianism. *K. Nederlandse Acad. Wetenschappen afd. Letterkunde, Med. (n.r.)*, **33**, 271–316.

Hörz, F. (1982) Ejecta of the Ries Crater, Germany in *Geological implications of impacts of large asteroids and comets on the Earth* (eds L.T. Silver and P.H. Schultz), Geol. Soc. Am. Spec. Pap., **190**, pp. 39–55.

Hsü, K.J. (1980) Terrestrial catastrophe caused by cometary impact at the end of Cretaceous. *Nature*, **285**, 201–3.

Hsü, K.J. (1982) Evolutionary and environmental consequences of a terminal Cretaceous event in *Cretaceous–Tertiary Extinctions and Possible Terrestrial and Extraterrestrial Causes* (eds D.A. Russell and G. Rice), Mus. of Canada, Syllogeus, **39**, pp. 140–42.

Hsü, K.J. (1984) Geochemical markers of impacts and their effects on environments in *Patterns of Change in Earth Evolution* (eds H.G. Holland and A.F. Trendall), Springer-Verlag, Berlin, pp. 63–74.

Hsü, K.J. (1986) Sedimentary petrology and biologic evolution. *J. Sedimentary Petrol.*, **56**, 729–32.

Hsü, K.J. *et al.* (1982) Mass mortality and its environmental and evolutionary consequences. *Science*, **216**, 249–56.

Hsü, K.J., McKenzie, J.A. and He, Q.X. (1982) Terminal Cretaceous environmental and evolutionary changes in *Geological implications of impacts of large asteroids and comets on the Earth* (eds L.T. Silver and P.H. Schultz), Geol. Soc. Am. Spec. Pap., **190**, pp. 317–28.

Hut, P. *et al.* (1987) Comet showers as a cause of mass extinctions. *Nature*, **329**, 118–26.

Hutton, J. (1788) *Theory of the Earth; or an investigation of the laws observable in the composition, dissolution, and restoration of land upon the globe.* R. Soc. Edinburgh, Trans., **1**(2), 209–304.

Hutton, J. (1975) *Theory of the Earth with Proofs and Illustrations.* Cadell and Davies, London; William Creech, Edinburgh, 2 vols.

Izett, G.A. and Bohor, B.F. (1986) Microstratigraphy of continental sedimentary rocks in the Cretaceous–Tertiary boundary interval in the Western Interior of North America. *Geol. Soc. Am. Abst. with Programs*, p. 644.

Izett, G.A. (1987) Authigenic 'spherules' in K–T boundary

sediments at Caravaca, Spain and Raton Basin, Colorado and New Mexico, may not be impact derived. *Geol. Soc. Am. Bull.*, **99**, 78–86.

Jablonski, D. (1985) Marine regressions and mass extinctions: a test using the modern biota in *Phanerozoic Diversity Patterns* (ed J.W. Valentine), Princeton University Press, Princeton, New Jersey, pp. 335–54.

Jablonski, D. (1986a) Background and mass extinctions: the alternations of macroevolutionary regimes. *Science*, **231**, 129–33.

Jablonski, D. (1986b) Causes and consequences of mass extinctions; a comparative approach in *Dynamics of Extinction* (ed D.K. Elliott), John Wiley and Sons, New York, pp. 183–229.

Jameson, R. (1976) *Elements of Geognosy. The Wernerian Theory of the Neptunian Origin of Rocks* (ed G.W. White), Hafner, New York.

Jansa, L.F. and Pe-Piper, G. (1987) Identification of an underwater extraterrestrial impact structure. *Nature*, **327**, 612–14.

Johnson, J.G. (1984) Comment on 'Temperature and biotic crises in the marine realm' by S.M. Stanley. *Geology*, **12**, 741.

Kalvoda, J. (1986) Upper Frasnian and Lower Tournaisian events and evolution of calcareous foraminifera – close links to climatic changes in *Global bio-events* (ed O.H. Walliser), *Lecture Notes in Earth Sciences*, **8**, Springer-Verlag, Berlin, p. 225.

Kauffman, E.G. (1984) The fabric of Cretaceous marine extinctions in *Catastrophes and Earth History* (eds W.A. Berggren and J.A. van Couvering), Princeton University Press, Princeton, New Jersey, pp. 151–246.

Kauffman, E.G. (1986) High-resolution event stratigraphy: regional and global Cretaceous bio-events in *Global Bio-events* (ed O.H. Walliser), *Lect. Notes on Earth Sciences*, **8**, Springer-Verlag, Berlin, pp. 279–335.

Kehew, A.E. and Lord, M.L. (1986) Origin and large-scale erosional features of glacial-lake spillways in the northern Great Plains. *Geol. Soc. Am. Bull.*, **97**, 162–77.

Keith, M.L. (1982) Violent volcanism, stagnant oceans, and some inferences regarding petroleum, strata-bound ores and mass extinctions. *Geochem. et Cosmochem. Acta*, **46**, 2621–37.

Keller, G. *et al.* (1983) Multiple microtektite horizons in Upper Eocene sediments: no evidence for mass extinctions. *Science*, **221**, 150–52.

Kelvin, W.T., First Baron (1852) On a universal tendency in nature to the dissipation of mechanical energy. *Phil. Mag.*, ser. 4, **4**, 304–6.

Kelvin, W.T., First Baron (1864) *On the secular cooling of the Earth*. R. Soc. Edinburgh, Trans., **23**, 157–70.
Kelvin, W.T., First Baron (1871) *On geological time*. Geol. Soc. Glasgow, Trans., **3**(1), 1–28.
Kelvin, W.T., First Baron (1899) The age of the Earth as an abode fitted for life. *Phil. Mag., ser. 5*, **47**, 69–90.
Kent, D.V. (1981) Asteroid extinction hypothesis. *Science*, **211**, 648–50.
Kerr, R.A. (1986a) Halley's confounding fireworks. *Science*, **234**, 1196–8.
Kerr, R.A. (1986b) Comets appear to be Rosetta stones. *Science*, **234**, 1321–2.
Kerr, R.A. (1987) Old and new geology meet in Phoenix. *Science*, **238**, 890.
Kirwan, R. (1793) *Examination of the supposed igneous origin of stony substances*. R. Irish Acad., Trans., **5**, 51–81.
Kirwan, R. (1799) *Geological Essays*. Printed by T. Bensley for D. Bremmer, London.
Knoll, A.H. (1984) Patterns of extinction in the fossil record of vascular plants in *Extinctions* (ed M.H. Nitecki), University of Chicago Press, Chicago, pp. 23–68.
Kollman, H.A. (1979) Distribution patterns and evolution of gastropods around the Cretaceous–Tertiary boundary in *Cretaceous–Tertiary Boundary Events* (eds W.K. Christensen and T. Birkelund), University of Copenhagen, Copenhagen, **2**, pp. 83–7.
Krassilov, V.A. (1975) Climatic changes in eastern Asia as indicated by fossil floras, II. Late Cretaceous and Danian. *Palaeogeog. Palaeoclimat. Palaeoecol.*, **17**, 157–72.
Krinov, E.L. (1963) Meteorite craters on the Earth's surface in *The Solar System* (eds B. Middlehurst and G.P. Kuiper), **4**, University of Chicago Press, Chicago, pp. 183–207.
Krinov, E.L. (1966) *Giant Meteorites* (trans. J.S. Romankiewicz), Pergamon Press, Oxford.
Kyte, F.T. and Wasson, J.T. (1986) Accretion rate of extraterrestrial matter; iridium deposited 33 to 67 million years ago. *Science*, **232**, 1225–9.
Kyte, F.T., Zhiming, Z. and Wasson, J.T. (1980) Siderophile-enriched sediments from the Cretaceous–Tertiary boundary. *Nature*, **288**, 651–6.
LaMont, C.C. (1943) Experiments on toleration of high temperature

in lizards with reference to adaptive coloration. *Ecology*, **24**, 94–108.

Laudan, R. (1987) *From Mineralogy to Geology: the Foundations of a Science, 1650–1830*, University of Chicago Press, Chicago and London.

Lerbekmo, J.F. *et al.* (1979) Magnetostratigraphy, biostratigraphy, and geochronology of Cretaceous–Tertiary boundary sediments, Red Deer Valley. *Nature*, **279**, 26–30.

Lerbekmo, J.F. *et al.* (1987) The relationship between the iridium anomaly and palynological floral levels at three Cretaceous–Tertiary boundary localities in western Canada. *Geol. Soc. Am. Bull.*, **99**, 325–30.

Lewis, J.S. *et al.* (1982) Chemical consequences of major impact events on Earth in *Geological implications of impacts of large asteroids and comets on the Earth* (eds L.T. Silver and P.H. Schultz), Geol. Soc. Am. Spec. Pap., **190**, pp. 215–21.

Lilly, P.A. (1981) Shock metamorphism in the Vredefort collar: evidence for internal shock sources. *J. Geophys. Res.*, **86**, 10689–700.

Lindinger, M. and Keller, G. (1987) Stable isotope stratigraphy across Cretaceous–Tertiary boundary in Tunisia: evidence for a multiple extinction mechanism? *Geol. Soc. Am. Abst. with Programs*, p. 747.

Lindsay, E.H., Butler, R.F. and Johnson, N.M. (1982) Testing of magnetostratigraphy in Late Cretaceous and Early Tertiary deposits, San Juan Basin, New Mexico in *Geological implications of impacts of large asteroids and comets on the Earth* (eds L.T. Silver and P.H. Schultz), Geol. Soc. Am. Spec. Pap., **190**, pp. 449–53.

Lutz, T.M. (1986) Evaluating periodic, episodic, and Poisson models of geologic time series: results for mass extinctions, magnetic reversals and meteorite impacts. *Geol. Soc. Am. Abst. with Programs*, p. 677.

Lyell, C. (1830–1833) *Principles of Geology*. John Murray, London, 3 vols.

Lyell, C. (1850) Anniversary address of the President. *Geol. Soc. London, Q. J.*, **6**, xxvii–lxvi.

Lyell, C. (1851) Anniversary address of the President. *Geol. Soc. London, Q. J.*, **7**, xxv–lxxvi.

McAlester, A.L. (1970) Animal extinctions, oxygen consumption, and atmospheric history, *J. Paleontology*, **44**, 405–9.

McCall, G.J.H. (1965a) New material from, and a reconsideration of, the Dalgaranga meteorite and crater, Western Australia. *Miner. Mag.*, **35**, 476–87.

McCall, G.J.H. (1965b) Possible meteorite craters – Wolf Creek, Australia and analogs. *New York Acad. Sci., Ann.*, **123**, 970–98.

McCall, G.J.H. (ed) (1977) *Meteorite Craters.* Dowden, Hutchinson, and Ross, Stroudsburg, Pennsylvania.

McCall, G.J.H. (ed) (1979) *Astroblems–Cryptoexplosion Structures.* Dowden, Hutchinson, and Ross, Stroudsburg, Pennsylvania.

McCrea, W.H. (1981) Long time-scale fluctuations in the evolution of the Earth, *R. Soc. London, Proc., ser. A.*, **375**, 1–41.

McGhee, G.R., Jr (1982) The Frasnian–Famennian extinction event: a preliminary analysis of Appalachian marine ecosystems in *Geological implications of impacts of large asteroids and comets on the Earth* (eds L.T. Silver and P.H. Schultz), Geol. Soc. Am. Spec. Pap., **190**, pp. 491–500.

McGhee, G.R., Jr, *et al.* (1986) Late Devonian 'Kellwasser event' mass extinction horizon in Germany: no geochemical evidence for a large-body impact. *Geology*, **14**, 776–9.

McKay, C.P. and Thomas, G.E. (1982) Formation of noctilucent clouds by an extraterrestrial impact in *Geological implications of impacts of large asteroids and comets on the Earth* (eds L.T. Silver and P.H. Schultz), Geol. Soc. Am. Spec. Pap., **190**, pp. 211–14.

McKinney, H.L. (1987) Taxonomic selectivity and continuous variation in mass and background extinctions of marine taxa. *Nature*, **325**, 143–5.

McLaren, D.J. (1970) Time, life, and boundaries. *J. Paleontology*, **44**, 801–15.

McLaren, D.J. (1982) Frasnian–Famennian extinctions in *Geological implications of impacts of large asteroids and comets on the Earth* (eds L.T. Silver and P.H. Schultz), Geol. Soc. Am. Spec. Pap., **190**, pp. 477–84.

McLaren, D.J. (1985) Mass extinction and iridium anomaly in the Upper Devonian of Western Australia: a commentary. *Geology*, **13**, 170–72.

McLean, D.M. (1978) A terminal Mesozoic 'greenhouse': lessons from the last. *Science*, **201**, 401–6.

McLean, D.M. (1981) A test of terminal Mesozoic 'catastrophe'. *Earth Planet. Sci. Letters*, **53**, 103–8.

McLean, D.M. (1982) Deccan volcanism and the Cretaceous–

Tertiary transition in *Cretaceous–Tertiary extinctions and possible terrestrial and extraterrestrial causes* (eds D.A. Russell and G. Rice), Mus. of Canada, Syllogeus, **39**, pp. 143–4.

McLean D.M. (1987) Volcanism induced trans-Cretaceous–Tertiary disorganization of the biosphere. *Geol. Soc. Am. Abst. with Programs*, pp. 767–8.

Macdougall, J.D. (1988) Seawater strontium isotopes, acid rain, and the Cretaceous–Tertiary boundary. *Science*, **239**, 485–7.

Madigan, C.T. (1937) The Boxhole Crater and the Huckitta meteorite (central Australia). *R. Soc. South Australia, Trans. and Proc.*, **61**, 187–90.

Malone, T.F. (1985) Preface in *Global Change* (eds T.F. Malone and J.G. Roederer), Cambridge University Press, London, pp. xi–xxi.

Mark, K. (1987) *Meteorite Craters*, University Arizona Press, Tucson.

Marvin, U. (1986) Meteorites, the moon and the history of geology. *J. Geol. Education*, **34**, 140–65.

Marx, J.L. (1978) Warm-blooded dinosaurs: evidence pro and con. *Science*, **199**, 1424–6.

Melosh, H.J. (1982) The mechanics of large meteoroid impacts in the Earth's oceans in *Geological implications of impacts of large asteroids and comets on the Earth* (eds L.T. Silver and P.H. Schultz), Geol. Soc. Am. Spec. Pap., **190**, pp. 121–7.

Melosh, H.J. (1985) Ejection of rock fragments from planetary bodies. *Geology*, **13**, 144–8.

Milton, D.J. (1972) *Structural Geology of the Henbury Meteorite Craters, Northern Territory, Australia.* U. S. Geol. Surv., Prof. Pap. 599-C, pp. C1–C17.

Montanari, A. *et al.* (1983) Spheroids at the Cretaceous–Tertiary boundary are altered impact droplets of basaltic composition. *Geology*, **11**, 668–71.

Mörner, N.-A. (1981) Revolution in Cretaceous sea-level analysis. *Geology*, **9**, 344–6.

Mörner, N.-A. (1984) Eustasy, geoid changes, and multiple geophysical reaction in *Catastrophes and Earth History* (eds W.A. Berggren and J.A. van Couvering), Princeton University Press, Princeton, New Jersey, pp. 395–415.

Muller, R.A. and Morris, D.C. (1986) Geomagnetic reversals from impacts on the Earth. *Geophys. Res. Letters*, **13**, 1177–80.

Murty, T.S. (1986) Comment on 'Protection of the human race against natural hazards (asteroids, comets, volcanoes, earthquakes)'. *Geology*, **14**, 633–4.

Mygatt, M. (1986) Dinosaur-find an earth shaker. *Dallas Morning News*, Aug. 9, 24A.

Napier, W.M. and Clube, S.V.M. (1979) A theory of terrestrial catastrophism. *Nature*, **282**, 455–9.

Naslund, H.R., Officer, C.B. and Johnson, G.D. (1986) Microspherules in Upper Cretaceous and Lower Tertiary clay layers at Gubbio, Italy. *Geology*, **14**, 923–6.

Newell, N.D. (1962) Paleontological gaps and geochronology. *J. Paleontology*, **36**, 592–610.

Newell, N.D. (1963) Crises in the history of life. *Scient. American*, **208**, 76–92.

Newell, N.D. (1967) Revolutions in the history of life in *Uniformity and simplicity, a symposium on the principle of uniformity of nature* (ed C.C. Albritton, Jr), Geol. Soc. Am. Spec. Pap., **89**, pp. 63–91.

Nichols, D.J. *et al.* (1986) Palynological and iridium anomalies at the Cretaceous–Tertiary boundary, south-central Saskatchewan. *Science*, **231**, 714–17.

Nunn, P.D. (1986) Implications of migrating geoid anomalies for the interpretation of high-level fossil coral reefs. *Geol. Soc. Am., Bull.*, **97**, 946–52.

Officer, C.B., Hallam, A., Drake, C.L. and Devine, J.D. (1987) Late Cretaceous and paroxysmal Cretaceous/Tertiary extinctions. *Nature*, **326**, 143–9.

O'Keefe, J.D. and Ahrens, T.J. (1982) The interaction of the Cretaceous/Tertiary bolide with the atmosphere, ocean, and solid Earth in *Geological implications of impacts of large asteroids and comets on the Earth* (eds L.T. Silver and P.H. Schultz), Geol. Soc. Am. Spec. Pap., **190**, pp. 103–20.

Olson, E.C. and Thomas, R.D.K. (1980) Introduction in *A cold look at the warm-blooded dinosaurs* (eds Olson and Thomas), Westview Press, Boulder, Colorado, pp. 1–14.

Olson, E.C. (1982) Extinctions of Permian and Triassic non-marine vertebrates in *Geological implications of impacts of large asteroids and comets on the Earth* (eds L.T. Silver and P.H. Schultz), Geol. Soc. Am. Spec. Pap., **190**, pp. 501–11.

Olson, P.E. *et al.* (1987) New Early Jurassic tetrapod assemblages constrain Triassic–Jurassic tetrapod extinction event. *Science*, **237**, 1025–9.

Orth, C.J. *et al.* (1981) An iridium abundance anomaly at the palynological Cretaceous–Tertiary boundary in northern New Mexico. *Science*, **214**, 1341–3.

Orth, C.J. *et al.* (1982) Iridium abundance measurements across the Cretaceous/Tertiary boundary in the San Juan and Raton basins of northern New Mexico in *Geological implications of impacts of large asteroids and comets on the Earth* (eds L.T. Silver and P.H. Schultz), Geol. Soc. Am. Spec. Pap., **190**, pp. 423–33.

Orth, C.J. *et al.* (1986) Terminal Ordovician extinction: geochemical analysis of the Ordovician/Silurian boundary, Anticosti Island, Quebec. *Geology*, **14**, 433–6.

Orth, C.J. *et al.* (1987) Iridium abundance peaks at Upper Cenomanian stepwise extinction horizons. *Geol. Soc. Am., Abst. with Programs*, p. 796.

Ostrom, J.H. (1980) The evidence for endothermy in dinosaurs in *A Cold Look at the Warm-blooded Dinosaurs* (eds E.C. Olson and R.D.K. Thomas), Westview Press, Boulder, Colorado, pp. 15–54.

Padian, K. (ed) (1986) *The Beginning of the Age of Dinosaurs*, Cambridge University Press, London.

Padian, K. (1987) Patterns of terrestrial faunal change across the Triassic–Jurassic boundary and the rise of the dinosaurs. *Geol. Soc. Am., Abst. with Programs*, p. 797.

Padian, K. *et al.* (1984) The possible influence of sudden events on biological relations and extinctions in *Patterns of Change in Earth Evolution* (eds H.D. Holland and A.F. Trendall), Springer-Verlag, Berlin, pp. 77–102.

Padian, K. and Clemens, W.A. (1985) Terrestrial vertebrate diversity: episodes and insights in *Phanerozoic Diversity Patterns* (ed J.W. Valentine), Princeton University Press, Princeton, New Jersey, pp. 41–96.

Perch–Nielsen, K., Ulteberg, K., and Evensen, J.E. (1979) Comments on 'The terminal Cretaceous event: a geologic problem with an oceanographic solution' in *Cretaceous–Tertiary Boundary Events* (eds W.K. Christensen and T. Birkelund), **2**, University of Copenhagen, Copenhagen, pp. 106–11.

Picard, M.D. (1985) The Cretaceous/Tertiary boundary controversy. *J. Geol. Education*, **33**, 80–83.

Pitman, W.C. III (1978) Relationship between eustasy and stratigraphic sequences of passive margins. *Geol. Soc. Am., Bull.*, **89**, 1389–1403.

Playfair, J. (1802) *Illustrations of the Huttonian Theory of the Earth*. Cadell and Davies, Edinburgh.

Playfair, J. (1805) *Biographical Account of the late James Hutton*, F.R.S., R. Soc. Edinburgh, Trans., **5**(3), 39–99.
Plinius Secundus, Gaius (1601) *The Historie of the World* (English translation of Pliny's *Natural History* by Philemon Holland), Adam Islep, London.
Pohl, J. *et al.* (1977) The Ries impact crater in *Impact and Explosion Cratering* (eds D.J. Roddy, R.O. Pepin and R.B. Merrill), Pergamon Press, New York, pp. 343–404.
Prinn, R.G. and Fegley, B., Jr (1987) Bolide impacts, acid rain, and biospheric traumas at the Cretaceous–Tertiary boundary. *Earth Planet. Sci. Letters*, **83**, 1–15.
Proctor, R.A. (1873) *The Moon; her motions, aspects, scenery, and physical condition*. Longmans, Green and Co., London.
Prosh, E.C. and McCracken, A.D. (1985) Postapocalypse stratigraphy: some considerations and proposals. *Geology*, **13**, 4–5.
Rampino, M. (1982) A non-catastrophist explanation for the iridium anomaly at the Cretaceous/Tertiary boundary in *Geological implications of impacts of large asteroids and comets on the Earth* (eds L.T. Silver and P.H. Schultz), Geol. Soc. Am. Spec. Pap., **190**, pp. 455–60.
Rampino, M.R. and Reynolds, R.C. (1983) Clay mineralogy of the Cretaceous–Tertiary boundary clay. *Science*, **219**, 495–8.
Rampino, M.R. and Stothers, R.B. (1984) Terrestrial mass extinctions, cometary impacts and the Sun's motion perpendicular to the galactic plane. *Nature*, **308**, 709–12.
Rampino, M.R. (1986) Volcanic winter? Atmospheric effects of the largest volcanic eruptions. *Geol. Soc. Am., Abst. with Programs*, p. 725.
Rampino, M.R. (1987) Impact cratering and flood basalt volcanism. *Nature*, **327**, 468.
Raup, D.M. (1982) Biogeographic extinctions: a feasibility test in *Geological implications of impacts of large asteroids and comets on the Earth* (eds L.T. Silver and P.H. Schultz), Geol. Soc. Am. Spec. Pap., **190**, pp. 277–81.
Raup, D.M. (1986a) *The Nemesis Affair*, W.W. Norton and Co., New York, London.
Raup, D.M. (1986b) Biological extinction in world history. *Science*, **231**, 1528–33.
Raup, D.M. and Sepkoski, J.J. Jr (1982) Mass extinctions in marine fossil record. *Science*, **215**, 1501–3.

Raup, D.M. and Sepkoski, J.J. Jr (1984) Periodicity of extinctions in the geologic past. *Natn. Acad. Sci., USA Proc.*, **81**, 801–5.

Raup, D.M. and Sepkoski, J.J. Jr (1986) Periodic extinctions of families and genera. *Science*, **231**, 833–6.

Raymond, A. *et al.* (1987) Comment and reply on 'Frasnian and Famennian mass extinctions and cold water oceans.' *Geology*, **15**, 777–8.

Reid, G.C. *et al.* (1978) Effects of intense stratospheric ionisation events. *Nature*, **275**, 489–92.

Reiff, W. (1977) The Steinheim Basin – an impact structure in *Impact and Explosion Cratering* (eds D.J. Roddy, R.O. Pepin and R.B. Merrill), Pergamon Press, New York, pp. 309–20.

Reinhardt, J. *et al.* (1986) Latest Cretaceous and earliest Tertiary sedimentation in the eastern Gulf Coastal Plain. *Geol. Soc. Am. Abst. with Programs*, p. 728.

Retallack, G.J. and Leahy, G.D. (1986) Cretaceous–Tertiary dinosaur extinction. *Science*, **234**, 1170–71.

Retallack, G.J., Leahy, G.D. and Spoon, M.D. (1987) Evidence from paleosols for ecosystem changes across the Cretaceous/Tertiary boundary in eastern Montana. *Geology*, **15**, 1090–93.

Rossiter, A.P. (1935) The first English geologist – Robert Hooke (1635–1703). *Durham Univ. J.*, **29**, 172–81.

Ruderman, M.A. (1974) Possible consequences of nearby supernova explosions for atmospheric ozone and terrestrial life. *Science*, **184**, 1079–81.

Rudwick, M.J.S. (1976) *The Meaning of Fossils; Episodes in the History of Paleontology*, 2nd edn, Neale Watson Academic Pubs., New York.

Russell, D.A. (1979) The enigma of the extinction of the dinosaurs. *Ann. Rev. Earth Planet. Sci.*, **7**, 163–82.

Russell, D.A. (1982) A paleontological consensus on the extinction of the dinosaurs in *Geological implications of impacts of large asteroids and comets on the Earth* (eds L.T. Silver and P.H. Schultz), Geol. Soc. Am. Spec. Pap., **190**, pp. 401–5.

Russell, D.A. (1984) The gradual decline of the dinosaurs – fact or fallacy. *Nature*, **307**, 360–1.

Russell, D.A. and Tucker, W. (1971) Supernovae and the extinction of the dinosaurs. *Nature*, **229**, 553–4.

Rutherford, E. (1904) The radiation and emanation of radium, pt. 2. *Technics*, Aug. issue, pp. 171–5.

Saito, T. and van Donk, J. (1974) Oxygen and carbon isotope measurements of Late Cretaceous and Early Tertiary foraminifera. *Micropaleontology*, **20**, 152–77.

Saito, T., Yamanoi, T. and Kaiho, K. (1986) End-Cretaceous devastation of terrestrial flora in the boreal Far East. *Nature*, **323**, 253–5.

Sanders, N.K. (1976) *The Epic of Gilgamesh. An English Version with an Introduction*. Penguin Books Ltd., Harmondsworth, England.

Saunders, W.B. and Ramsbottom, W.H.C. (1986) The mid-Carboniferous eustatic event. *Geology*, **14**, 208–12.

Sawatzky, H.B. (1977) Buried impact craters in the Williston Basin and adjacent area in *Impact and Explosion Craters* (eds D.J. Roddy, R.O. Pepin and R.B. Merrill), Pergamon Press, New York, pp. 461–80.

Schatz, A. (1957) Some biochemical and physiological considerations regarding the extinction of the dinosaurs. *Pennsylvania Acad. Sci., Proc.*, **31**, 26–36.

Schindewolf, O.H. (1963) Neokatastrophismus? *Deut. Geol. Ges. Z.*, **114**, 430–45.

Schindewolf, O.H. (1977) Neocatastrophism? (English translation of 1963 article), *Catastrophist Geol.*, **2**, 9–21.

Schleich, H.H. (1986) Reflections upon the changes of local Tertiary herpetofaunas to global events in *Global Bio-events* (ed D.H. Walliser), Lect. Notes in Earth Sci. 8, Springer-Verlag, Berlin, pp. 429–42.

Schopf, T.J.M. (1974) Permo-Triassic extinctions: relation to sea-floor spreading. *J. Geol.*, **82**, 129–43.

Schopf, T.J.M. (1982) Extinction of the dinosaurs: a 1982 understanding in *Geological implications of impacts of large asteroids and comets on the Earth* (eds L.T. Silver and P.H. Schultz), Geol. Soc. Am. Spec. Pap., **190**, pp. 415–22.

Schopf, T.J.M. *et al.* (1971) Oxygen consumption rates and their paleontologic significance. *J. Paleontology*, **45**, 247–52.

Schwartz, R.D. and James, P.B. (1984) Periodic mass extinctions and the sun's oscillation about the galactic plane. *Nature*, **308**, 712–13.

Schwartzschild, B. (1987) Do asteroid impacts trigger geomagnetic reversals? *Physics Today*, **40**(2), 17–20.

Schweizer, F. (1986) Colliding and merging galaxies. *Science*, **231**, 227–34.

Sedgwick, A. (1831) Address to the Geological Society, delivered on the evening of the 18th February. *Geol. Soc. London, Proc.*, **1**, 281–316.

Sepkoski, J.J. Jr (1982) Mass extinctions in the Phanerozoic oceans: a review in *Geological implications of impacts of large asteroids and comets on the Earth* (eds L.T. Silver and P.H. Schultz), Geol. Soc. Am. Spec. Pap., **190**, pp. 283–9.

Sepkoski, J.J. Jr (1986a) Phanerozoic overview of mass extinction in *Patterns and Processes in the History of Life* (eds D.M. Raup and D. Jablonski), Springer-Verlag, Berlin, pp. 277–95.

Sepkoski, J.J. Jr (1986b) Global bio-events and the question of periodicity in *Global Bio-events* (ed O.H. Walliser), Lect. Notes in Earth Sciences 8, Springer-Verlag, Berlin, pp. 47–61.

Sepkoski, J.J Jr and Raup, D.M. (1986) Periodicity in marine extinction events in *Dynamics of Extinction* (ed D.K. Elliott), John Wiley and Sons, New York, pp. 1–36.

Shah, B.V. (1983) Is the environment becoming more hazardous? A global survey, 1947 to 1980. *Disasters*, **7**, 202–9.

Sheehan, P.M. and Morse, C.L. (1986) Cretaceous–Tertiary dinosaur extinction. *Science*, **234**, 1171–2.

Shoemaker, E.M. (1960) Penetration mechanics of high-velocity meteorites, illustrated by Meteor Crater, Arizona. *21 Internat. Geol. Congr., Copenhagen, Proc., sec. 18*, pp. 418–34.

Shoemaker, E.M. (1977) Why study impact craters? in *Impact and Explosion Cratering* (eds D.J. Roddy, R.O. Pepin and R.B. Merrill), Pergamon Press, New York, pp. 1–10.

Shoemaker, E.M. (1983) Asteroid and comet bombardment of the Earth. *Earth Planet. Sci., Ann. Rev.*, **11**, 461–94.

Shoemaker, E.M. (1986) Geologic history of the Uranian satellites. *Geol. Soc. Am. Abst. with Programs*, p. 749.

Shoemaker, E.M. *et al.* (1963) Hypervelocity impact of steel into Coconino Sandstone. *Am. J. Sci.*, **261**, 668–82.

Shoemaker, E.M. and Kieffer, S.W. (1974) *Guidebook to the Geology of Meteor Crater, Arizona*. Arizona State University, Center for Meteorite Stud., Pub. 17.

Shoemaker, E.M. and Shoemaker, C.S. (1987) Meteorite craters of Western Australia. *Geol. Soc. Am. Abst. with Programs*, pp. 842–3.

Signor, P.W. and Lipps, J.H. (1982) Sampling patterns, gradual extinction patterns and catastrophes in the fossil record in

Geological implications of impacts of large asteroids and comets on the Earth (eds L.T. Silver and P.H. Schultz), Geol. Soc. Am. Spec. Pap., **190**, pp. 291–6.

Silver, L.T., McLaren, D.J. and Drake, C.L. (1982) Impacts and evolution conference: reports and comments. *Geology*, **10**, 126–8.

Simpson, G.G. (1970) Uniformitarianism: an inquiry into principles, theory, and method in geohistory and biohistory in *Essays in Evolution and Genetics in Honor of Theodosius Dobzhansky* (eds M.K. Hecht and W.C. Steere), Appleton-Century-Crofts, New York, pp. 43–96.

Sloan, R.E. *et al.* (1986) Gradual dinosaur extinction and simultaneous ungulate radiation in the Hell Creek Formation. *Science*, **232**, 629–33.

Sloan, R.E. (1987) Paleocene dinosaur extinction in South China. *Geol. Soc. Am., Abst. with Programs*, p. 848.

Smit, J. (1982) Extinction and evolution of planktonic foraminifera after a major impact at the Cretaceous/Tertiary boundary in *Geological implications of impacts of large asteroids and comets on the Earth* (eds L.T. Silver and P.H. Schultz), Geol. Soc. Am. Spec. Pap., **190**, pp. 329–52.

Smit, J. and Hertogen, J. (1980) An extraterrestrial event at the Cretaceous–Tertiary boundary. *Nature*, **285**, 198–200.

Smit, J. and Klaver, G. (1981) Sanadine spherules at the Cretaceous–Tertiary boundary indicate a large impact event. *Nature*, **292**, 47–9.

Smith, W. (1815) *A Memoir to the Map and Delineation of the Strata of England and Wales, with part of Scotland.* Printed for John Carey, London.

Spencer, L.J. (1933) Meteoritic iron and silica-glass from the meteorite craters of Henbury (Central Australia) and Wabar (Arabia). *Mineral. Mag.*, **23**, 387–404.

Stanley, S.M. (1984a) Temperature and biotic crises in the marine realm. *Geology*, **12**, 205–8.

Stanley, S.M. (1984b) Mass extinctions in the ocean. *Scient. American*, **250**(6), 64–72.

Stearn, C.W. (1987) Effect of the Frasnian–Famennian extinction event upon the stromatoporoids. *Geology*, **15**, 677–9.

Steckler, M. (1984) Changes in sea-level in *Patterns of Change in Earth*

Evolution (eds H.D. Holland and A.F. Trendall), Springer-Verlag, Berlin, pp. 103–21.

Steno, N. (1669) *De Solido Intra Solidum Naturaliter Contento. Dissertationis Prodromus*. typ. sub signo Stellae, Florence.

Steno, N. (1968) *The Prodromus of Nicolaus Steno's Dissertation Concerning a Solid Body Enclosed by Process of Nature Within a Solid* (English translation of original Latin text by J.G. Winter), Hafner, New York and London.

Stevens, C.H. (1977) Was development of brackish oceans a factor in Permian extinctions? *Geol. Soc. Am., Bull.*, **88**, 133–8.

Stevenson, D.J. (1986) Lunar origin. *Science*, **234**, 1016–17.

Stothers, R.B. (1984) The great Tambora eruption of 1815 and its aftermath. *Science*, **224**, 1191–8.

Sullivan, R.M. (1987) Dinosaurs, asteroids and extinction: reptilian diversity across the Cretaceous–Tertiary boundary reassessed. *Geol. Soc. Am. Abst. with Programs*, p. 859.

Surlyk, F. (1980) The Cretaceous–Tertiary boundary event. *Nature*, **285**, 187–8.

Tappan, H. (1982) Extinction or survival: selectivity and causes of Phanerozoic crises in *Geological implications of impacts of large asteroids and comets on the Earth* (eds L.T. Silver and P.H. Schultz), Geol. Soc. Am. Spec. Pap., **190**, pp. 265–76.

Tappan, H. (1986) Phytoplankton: below the salt at the global table. *J. Paleontology*, **60**, 545–54.

Taylor, S.R. (1965) The Wolf Creek iron meteorite. *Nature*, **208**, 944–5.

Terry, K.D. and Tucker, W.H. (1968) Biological effects of supernovae. *Science*, **159**, 421–3.

Thierstein, H. (1979) The terminal Cretaceous oceanic event in *Cretaceous–Tertiary Boundary Events* (eds W.K. Christensen and T. Birkelung), University of Copenhagen, Copenhagen, **2**, pp. 22–8.

Thierstein, H. and Berger, W.H. (1978) Injection events in ocean history. *Nature*, **276**, 461–6.

Thomson, K.T. (1988) Anatomy of the extinction debate. *Am. Scientist*, Jan–Feb., **76**, 59–61.

Toon, O.B. (1984) Sudden changes in atmospheric composition and climate in *Patterns of Change in Earth Evolution* (eds H.D. Holland and A.F. Trendall), Springer-Verlag, Berlin, pp. 41–61.

Trefil, J.S. and Raup, D.M. (1987) Numerical simulations and the problem of periodicity in the cratering record. *Earth Planet. Sci. Letters*, **82**, 159–64.

Tschudy, R.H. (1984) Palynological evidence for change in continental floras at the Cretaceous–Tertiary boundary in *Catastrophies and Earth History* (eds W.A. Berggren and J.A. van Couvering), Princeton University Press, Princeton, New Jersey, pp. 315–37.

Tschudy, R.H. and Tschudy, B.D. (1986) Extinction and survival of plant life following the Cretaceous/Tertiary boundary event, Western Interior, North America. *Geology*, **14**, 667–70.

Turco, R.P. *et al.* (1981) Tungushka meteor fall of 1908: effects on stratospheric ozone. *Science*, **214**, 19–23.

Uffen, R.J. (1963) Influence of the earth's core on the origin and evolution of life. *Nature*, **198**, 143–4.

Upchurch, G.R. Jr (1987) Plant extinction patterns at the Cretaceous–Tertiary boundary, Raton and Denver basins. *Geol. Soc. Am., Abst. with Programs*, p. 874.

Urey, H.C. (1973) Cometary collisions and geological periods. *Nature*, **242**, 32–3.

Vail, P.R. *et al.* (1977) Seismic stratigraphy and global changes of sea-level in *Seismic Stratigraphy: Applications to Hydrocarbon Exploration, Pt. 4* (ed C.E. Payton), Am. Ass. Petroleum Geologists, Mem. 26, pp. 83–97.

Valentine, J.W. and Moores, E.M. (1970) Plate-tectonic regulation of faunal diversity and sea-level: a model. *Nature*, **228**, 657–9.

Van den Bergh, S. (1981) Size and age of the universe. *Science*, **213**, 825–30.

Van Valen, L.M. and Sloan, R.E. (1977) Ecology and the extinction of the dinosaurs. *Evolutionary Theory*, **2**, 37–64.

Van Valen, L.M. (1984) Catastrophes, expectations, and the evidence. *Paleobiology*, **10**, 121–37.

Vickery, A.M. and Melosh, H.J. (1987) The large crater origin of the SNC meteorites. *Science*, **237**, 738–43.

Vitaliano, D.B. (1973) *Legends of the Earth. Their Geologic Origins*, Indiana University Press, Bloomington, Indiana.

Vogt, P.R. and Holden, J.C. (1979) The end-Cretaceous extinctions; a study of the multiple working hypotheses gone mad in *Cretaceous–Tertiary Boundary Events* (eds W.K. Christensen and T. Birkelund), University of Copenhagen, **2**, Copenhagen, p. 49.

Waldrop, M.M. (1986) Voyage to a blue planet. *Science*, **231**, 916–18.

Wasson, J.T. (1985) *Meteorites: Their Record of Early Solar-System History*, W.H. Freeman and Co., New York.

Weaver, K.F. (1986) Meteorites: invaders from space. *Natn. Geographic*, **170**, 390–418.

Weissman, P.R. (1982) Terrestrial impact rates for long and short-period comets in *Geological implications of impacts of large asteroids and comets on the Earth* (eds L.T. Silver and P.H. Schultz), Geol. Soc. Am. Spec. Pap., **190**, pp. 15–24.

Werner, A.G. (1774) *On the External Characters of Minerals* (tr. A.V. Carozzi), University of Illinois Press, Urbana.

Werner, A.G. (1786) *Short Classification and Description of the Various Rocks* (tr. A. Ospovat), Hafner, New York.

Werner, E. (1904) Das Ries in der Schwabischen-Frankischen Alb. *Bl. Schwab. Albvieren*, **16**, 153–67.

Wetherill, G.W. and Shoemaker, E.M. (1982) Collision of astronomically observable bodies with the Earth in *Geological implications of impacts of large asteroids and comets on the Earth* (eds L.T. Silver and P.H. Schultz), Geol. Soc. Am. Spec. Pap., **190**, pp. 1–13.

Whewell, W. (1832) Review of Lyell's *Principles of Geology*, **2**, Q. *Rev.*, March, 1832, **93**, 103–32.

Whewell, W. (1872) *History of the Inductive Sciences from the Earliest to the Present Time*, 3rd edn, 2 vols, D. Appleton and Co., New York.

Whiston, W. (1696) *A New Theory of the Earth, from its Original, to the Consummation of all Things*, R. Roberts, London.

Whitmire, D.P. and Jackson, A.A. IV (1984) Are periodic mass extinctions driven by a distant solar companion? *Nature*, **308**, 713–15.

Whitmire, D.P. and Matese, J.J. (1985) Periodic comet showers and Planet X. *Nature*, **313**, 36–8.

Whittow, J. (1979) *Disasters: the Anatomy of Environmental Hazards*, University of Georgia Press, Athens, Georgia.

Whyte, M.A. (1977) Turning points in Phanerozoic history. *Nature*, **267**, 679–82.

Wiedmann, J. (1986) Macro-invertebrates and the Cretaceous–Tertiary boundary in *Global Bio-events* (ed O.H. Walliser), Lect. Notes in Earth Sci. 8, Springer-Verlag, Berlin, pp. 397–409.

Wilde, P. *et al.* (1986) Iridium abundances across Ordovician–Silurian stratotype. *Science*, **233**, 339–41.

Wilford, J.N. (1985) *The Riddle of the Dinosaur*, Knopf, New York.

Williams, G.E. (1986) The Acraman impact structure: source of ejecta in Late Precambrian shales, South Australia. *Science*, **233**, 200–203.

Wilshire, H.G. *et al.* (1972) Geology of the Sierra Madera cryptoexplosion structure, Pecos County, Texas, *US Geol. Survey, Prof. Pap.*, 599-H.

Wilson, L.G. (1972) *Charles Lyell. The years to 1841: the Revolution in Geology*, Yale Univ. Press, New Haven and London.

Wolbach, W.S., Lewis, R.S. and Anders, E. (1985) Cretaceous extinctions: evidence for wildfires and search for meteoritic material. *Science*, **230**, 167–70.

Woodwell, G.M. (1963) The ecological effects of radiation. *Scient. American*, **208**, 40–49.

Woodwell, G.M. (1967) Radiation and the patterns of Nature. *Science*, **156**, 461–70.

Worsley, T.R. (1971) Terminal Cretaceous events. *Nature*, **230**, 318–20.

Wylie, C.C. (1934) Meteorite craters, meteors and bullets. *Pop. Astron.*, **41**, 211–14.

Zhou, L. (1987) Trace-element geochemistry of the Permian–Triassic boundary. *Geol. Soc. Am., Abst. with Programs*, p. 904.

Zinsmeister, W.J. *et al.* (1987) Faunal transitions across the K/T boundary in Antarctica. *Geol. Soc. Am., Abst. with Programs*, p. 906.

von Zittel, K. (1901) *History of Geology and Palaeontology to the end of the nineteenth century* (tr. M.M. Ogilvie-Gordon), Walter Scott, London.

Zoller, W.H. *et al.* (1983) Iridium enrichment in airborne particles from Kiluea Volcano: January, 1983. *Science*, **222**, 1118–21.

Index